Health Effects of Metals and Related Substances in Drinking Water

Health Effects of Metals and Related Substances in Drinking Water

Authors

Prof. M. Ferrante, Dr. G. Oliveri Conti, Prof. Z. Rasic-Milutinovic and Dr. D. Jovanovic

Coauthors

Giovanni Arena University of Catania, (IT)

Chiara Copat University of Catania, (IT)

Maria Cunsolo University of Catania, (IT)

Maria Grazia D'Agati University of Catania, (IT)

Adriana Floridia University of Catania, (IT)

Maria Fiore University of Catania, (IT)

Roberto Furnari University of Catania,(IT)

Caterina Ledda University of Catania, (IT)

Ignatius C. Maduka University of Nigeria, (WAN)

Cristina Mauceri University of Catania, (IT)

Carlotta Malagoli University of Modena and Reggio Emilia, (IT)

Sanjay Mishra, IFTM Campus, (IND)

Michael R. Moore, Griffith University, (AUS)

Marco Vinceti University of Modena and Reggio Emilia, (IT)

Published by **IWA Publishing**
Alliance House
12 Caxton Street
London SW1H 0QS, UK
Telephone: +44 (0)20 7654 5500
Fax: +44 (0)20 7654 5555
Email: publications@iwap.co.uk
Web: www.iwapublishing.com

First published 2014

British Library Cataloguing in Publication Data
A CIP catalogue record for this book is available from the British Library

ISBN: 9781780405971 (Paperback)
ISBN: 9781780405988 (eBook)

Contents

Contents

Contents

Contents

About The Authors

Editors & Authors

Prof. Margherita Ferrante
MD and Biologist, hygiene specialist and general pathology specialist, Associate Professor of General and Applied Hygiene of Medicine School and Sciences School of Catania University, Director of Environmental Hygiene and Food Laboratories (LIAA) of "G.F. Ingrassia" Department - Hygiene and Public Health, University of Catania, Italy.
Coordinator of the "Health and Environment" Committee of the Italian Society of Hygiene and Preventive Medicine.
Coordinator of "Master enabler for occupational medicine"
Component of scientific Committee of regional master "Monitoring and evaluation of mutagenic, carcinogenic and teratogenic environmental risk."
Member of "Metals and Related Substances in Drinking Water" IWA Specialist Group.
Research Topics: public health, environmental hygiene, ecotoxicology, food hygiene, behaviour and related diseases.

Dr. Gea Oliveri Conti
B.Sc., PhD. Researcher of General and Applied Hygiene of Medicine School and Sciences School of Catania University. Researcher of Environmental Hygiene and Food Laboratories. Hygiene Adjunct Professor of Political and Social Sciences Department and "G.F. Ingrassia" Department of Hygiene and Public Health, Catania University.
Research Topics: Environmental hygiene, ecotoxicology, food hygiene, public health, behaviour and related diseases.

Prof. Zorica Rasic-Milutinovic
MD, PhD. Head of Department of Endocrinology, Clinic of Internal Medicine, University Hospital Zemun/Belgrade, University of Belgrade, Serbia. Professor of Research, Institute for Medical Research, Center of Excellence for metabolism and nutrition, University of Belgrade. Professor in US Medical School, Belgrade, Serbia.

Dr. Dragana Jovanovic
MD, M.Sc, hygiene specialist. Institute of Public Health "Dr Milan Jovanovic Batut", Belgrade, Center for Hygiene and Human Ecology, Department of Drinking water and recreational water quality, of Serbia
Research Topics: Drinking and recreational water quality. Drinking water arsenic and its health impact.

Coauthors

Giovanni Arena
M.Sc. Specialist in heavy metals analysis. Department "G.F. Ingrassia" - Hygiene and Public Health, University of Catania, Italy.

Chiara Copat
M.Sc. PhD. Specialist in heavy metals analysis. Department "G.F. Ingrassia" - Hygiene and Public Health, University of Catania, Italy.

Maria Cunsolo
MD. Department "G.F. Ingrassia" - Hygiene and Public Health, University of Catania, Italy.

Maria Grazia D'Agati
MD. Department "G.F. Ingrassia" - Hygiene and Public Health, University of Catania, Italy.

Dr. Adriana Floridia.
MD. Department "G.F. Ingrassia" - Hygiene and Public Health, University of Catania, Italy.

Prof. Maria Fiore
MD. PhD, Hygiene specialist, Researcher, Department "G.F. Ingrassia" - Hygiene and Public Health, University of Catania, Italy.

Roberto Furnari
MD. Department "G.F. Ingrassia" - Hygiene and Public Health, University of Catania, Italy.

Caterina Ledda
MLS, Department "G.F. Ingrassia" - Hygiene and Public Health, University of Catania, Italy.

Ignatius C. Maduka
MD. PhD. Department of Chemical Pathology, University of Nigeria Teaching Hospital (UNTH), Ituku-Ozalla, Enugu, Nigeria.

Cristina Mauceri
MD. Department "G.F. Ingrassia" - Hygiene and Public Health, University of Catania, Italy.

Carlotta Malagoli
MD. PhD. Researcher. CREAGEN - Environmental, Genetic and Nutritional Epidemiology Research Center, Department of Diagnostic and Clinical Medicine and of Public Health, University of Modena and Reggio Emilia, Modena, Italy.

Sanjay Mishra
Professor & Dean School of Biotechnology IFTM University Lodhipur Rajput, Delhi Road (NH- 24) Moradabad 244 102, U.P. India.

Michael R. Moore
B.Sc.,Ph.D., D.Sc. Biochemistry in Medicine, Toxicology, Chemistry. Chair, Water Quality Research Australia (WQRA). Vice-president Australasian College of Toxicology & Risk Assessment. Emeritus Professor The University of Queensland. Honorary Professor Griffith University.

Marco Vinceti
Associate Professor of General and Applied Hygiene, CREAGEN - Environmental, Genetic and Nutritional Epidemiology Research Center, Department of Diagnostic and Clinical Medicine and of Public Health, University of Modena and Reggio Emilia, Modena, Italy.

Review panel

Michelle Giddings, Michael Moore and Robert Bos make up the Review Panel.

Foreword

In December 2006, an international research network on “metals and related substances in drinking water” was established with funding from COST (www.cost.eu), an institution within the European Union. Over its four year life-span it built up representation from 27 European countries and established links with Canada, the US, the European Commission’s Joint Research Centre and the World Health Organization. It was successful in convening four international conferences in which the occurrence and impact of a wide range of metals and metalloids in drinking water were discussed. These conferences enabled the results of numerous research and practitioner experiences to be assimilated, creating a much better understanding of the extent and nature of the problems from metals and metalloids in drinking water, in both Europe and North America.

In November 2010, this international research network became a Specialist Group within the International Water Association and its members now come from all over the world. The Specialist Group continues to be active in the topic of metals and related substances in drinking water and has published a range of Best Practice Guides, Codes of Practice and Research Reports (www.iwap.co.uk).

Many of the metals and metalloids that are found in drinking water can have an adverse impact on human health. This book provides a “state-of-the-art” review of the health implications of metals and metalloids in drinking water and will be a key reference in the risk assessment and management of water supplies. The book is the second published by the Specialist Group in its Research Report Series.

Dr. Colin Hayes
Chairman
IWA Specialist Group on Metals and Related Substances in Drinking Water

Preface

Margherita Ferrante.

Water is the basic constituent (see Table 1) of all living beings it is, therefore, an essential dietary element (see Table 2) and a primary resource. The International standard references concerning water resources are various and, though they are based on WHO guidelines, they are extremely diversified in relation to local issues and emerging problems.

Table 1. Water's effect on the body	
Human body composition (65 - 90 %)	(Sciacca & Oliveri Conti, 2009; Popkin et al., 2010)
Thermoregulation	(Sawka et al., 2005; Sciacca & Oliveri Conti, 2009)
Lubrication	(Sciacca & Oliveri Conti, 2009; Popkin et al., 2010)
Solvent effect	(Sciacca & Oliveri Conti, 2009; Popkin et al., 2010)
Catalytic function in biochemical reactions	(Sawka,et al., 2005; Popkin et al., 2010)
Hydration	(Sciacca & Oliveri Conti, 2009; Popkin et al., 2010)
Digestion	(Sciacca & Oliveri Conti, 2009; Popkin et al., 2010)
Nutrient transport	(Sciacca & Oliveri Conti, 2009; Popkin et al., 2010)
Absorption of nutrients	(Sciacca & Oliveri Conti, 2009; Popkin et al., 2010)
Removing waste by urine	(Sciacca & Oliveri Conti, 2009; Yosef & Shalaby, 2010)

Table 2.Tissues and organs with high water content		
Amniotic liquid	(99%)	(Bacchi Modena & Fieni, 2004)
Brain	(75%)	(Nieuwenhuys et al., 1998)
Bone marrow	(99%)	(Sciacca & Oliveri Conti, 2009)
Blood plasma	(85%)	(Krebs, 1950)
Breast milk	(88%)	(Marvulli, 2010)
Cerebrospinal fluid	(99%)	(Bulat et al., 2008)
Vitreous humor	(99%)	(Nickerson , 2006)

Metals are elements that occur naturally in geological formations. Naturally occurring metalscan dissolve in water when it comes into contact with rocks or soil. Some metals are essential for life and are naturally available in our food and water. Trace amounts of metals are common in water, and these are normally not harmful to your health. In fact, some metals are essential to sustain life. Calcium, magnesium, potassium, and sodium must be present for normal body functions. Cobalt, copper, iron, manganese, molybdenum, selenium, and zinc, are needed at low levels as catalysts and co-factors for enzyme activities (Sciacca & Oliveri Conti; 2009; Sciacca et al., 2011).

Increased urbanization and increased water demand in industrial areas has caused issues with metals in groundwaters. In fact the contamination of our water resources by poisonous metals occurs largely due to human activity. These activities include industrial processes, such as electronics industry and the mining activity, the agricultural activities, and the dumping of wastes in landfills. Though in small quantities, some metals are nutritionally essential for a healthy life, abnormal amounts of any of them may cause acute or chronic toxicity (poisoning) and even cancer due to long-term oral exposures. Drinking water containing metals such as aluminum, arsenic, barium, cadmium, chromium, lead, mercury, selenium, and silver, may be hazardous to health. Trace amounts of metals enter our water supplies naturally as rain and percolates through rock. This water enters larger water bodies that we then use as resources for drinking water (Sciacca & Oliveri Conti; 2009; Sciacca et al., 2011).

It is really important in the light of these considerations to protect all water reserves used as sources of potable water supply from any forms of contamination. Knowledge of the chemical and toxicological properties of metals makes it possible to provide an efficient and effective prevention of waterborne disease caused by metals contamination.

References:

- Bacchi Modena A., Fieni S. (2004). Amniotic fluid dynamics. Conference report. *Acta Bio Medica Ateneo Parmense*; 75(1): 11-13. http://www.actabiomedica.it/data/2004/supp_1_2004/bacchi_2.pdf
- Bulat M., Lupret V., Orehković D., Klarica M. (2008). Transventricular and transpial absorption of cerebrospinal fluid into cerebral microvessels. *Coll Antropol.*; **32**(1):43-50.
- Kavouras S.A., Anastasiou C.A. (2010). Water Physiology: Essentiality, Metabolism, and Health Implications. *Nutrition Today.* **45**(6):S27-32.
- Krebs H.A. (1950). Chemical Composition of Blood Plasma and Serum. *Annual Review of Biochemistry*, **19**:409-30.
- Marvulli L. (2010). The resources of human milk and its treatment. Doctoral thesis. Doctorate in maternal-child and in developmental age medicine and pathophysiology of sexual reproduction. University of Bologna.http: unibo.it
- Nickerson C.S. (2006). Thesis. Chapter 3: The vitreous humor: mechanics and structure. http://thesis.library.caltech.edu/974/3/CSN_CH3.pdf
- Nieuwenhuys R., Ten Donkelaar H.J., Nicholson C. (1998). *The Central Nervous System of Vertebrates*. Vol. 3, Berlin: Springer.
- Popkin B.M., D'Anci K.E., Rosenberg I.H. (2010). Water, Hydration and Health. *Nutr Rev.*; **68**(8): 439–58.
- Sawka MN, Cheuvront SN, Carter R. (2005). 3rd Human water needs. *Nutr Rev*, **63**:S30–39.
- Sciacca S., Oliveri Conti G. (2009). Mutagens and carcinogens in drinking water. *Mediterranean Journal Of Nutrition and Metabolism*, **2**:157-62.
- Sciacca S., Ferrante M., Oliveri Conti G. (2011). Mutagens and Carcinogens in Water Resources. Nova Science Publisher Inc. 400 Oser Avenue, Suite 1600, Hauppauge New York. ISBN: 978-1-61324-599-6 2011.
- Shea D. (1988). Developing National Sediment Quality Criteria. *Environ. Sci. and Tech.*, **22**(11):1257-61.

- United States Environmental Protection Agency (USEPA). (1999). National Recommended Water. Quality Criteria. EPA 822-Z-99-001. Washington, D.C.
- United States Environmental Protection Agency (USEPA). (1998). National primary drinking water regulations: disinfectants and disinfection by-products rule. FedReg 63 (241),69389–476.
- Yosef E.S.M., Shalaby M.N. (2010). Effect of a Nutrition Compound (Honey and Water) on Blood Glucose, Body Temperature and Some Physiological Variables in Wrestlers. *World J. Sport Sci.*, **3**(S):930-5.

Aknowledgements

We wish to thank all authors and coauthors for precious suggestions and for meticulous work carried out for draftiong of this monograph.

Thanks al*so to the precious* collaboration of Dr. Pasquale Di Mattia (MD, hygiene specialist) for helping in critically reviewing the English grammar of the monograph.

This book is dedicated to all our children.

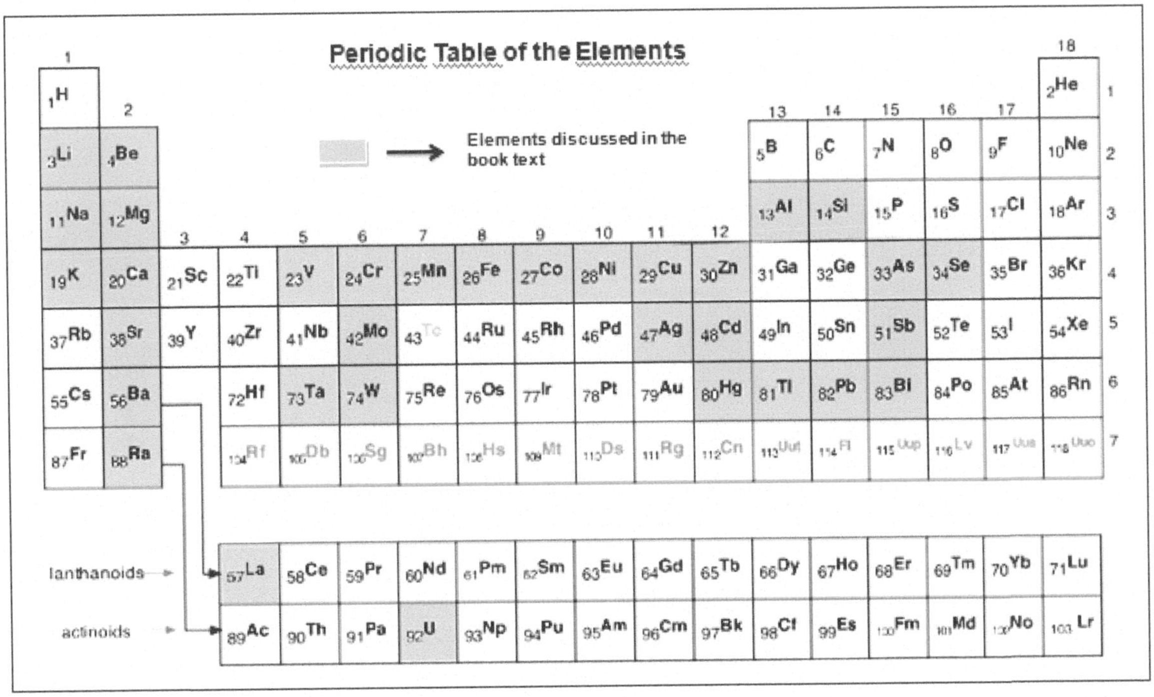

Chapter 1
Metals and drinking water.

Margherita Ferrante, Gea Oliveri Conti.

Drinking water quality is of direct relevance to human health. It reflects the levels of contaminants in the surface water and groundwater, and the efficiency of water treatment and distribution.

Many heavy metals belong to the so-called “trace elements”, but in recent decades the flow from the hydrosphere to man for metalloids and metals such as Arsenic, Lead, Mercury, Chromium, etc..has increased substantially due to release from industrial processes, use as pesticides and natural releases from soil into groundwater (Sciacca & Oliveri Conti, 2009; Sciacca et al., 2011). In fact, heavy metals have been used by humans for thousands years. Although adverse health effects of heavy metals have been known for a long time, exposure to heavy metals continues, and is even increasing in some parts of the world, particularly in less developed countries. Some metals are essential for life and are naturally available in our food and water (Deveau, 2010). Toxic metals are elements and represent the ultimate form of persistent environmental pollutants because they are chemically and biologically indestructible.

Health concerns associated with heavy metals in drinking water may arise from massive accidental contamination of a drinking water supply, but exposure comes primarily from prolonged periods of chronic exposure to trace doses of the metals; as a consequence, understanding the relationship between drinking-water quality and disease is very important. Pesticide and metal contamination of drinking water supplies has been identified as problem in many European countries. The most common problem across the European Community countries is metal contamination. The Czech Republic has problems with barium, nickel and selenium and in Lithuania 55% of drinking water sources have an excess of iron. Problems with high doses of iron and manganese in tap water are common in Central and Eastern European countries due to lack of efficient technologies for removal of these contaminants that often occur naturally in groundwater. In addition, Slovakia and Hungary have high concentrations of arsenic (see Fig.1.1), a toxic metalloid (Trent & Thyssen, 2003). Sources of arsenic in drinking water are from waters flowing through arsenic rich rocks but also from industrial contamination (Sciacca & Oliveri Conti, 2009; Trent & Thyssen, 2003).

In conclusion, alhough emissions have declined in most developed countries over the last 100 years, in addition to metals essential for life, drinking water may still contain metals that cause chronic or acute poisoning.

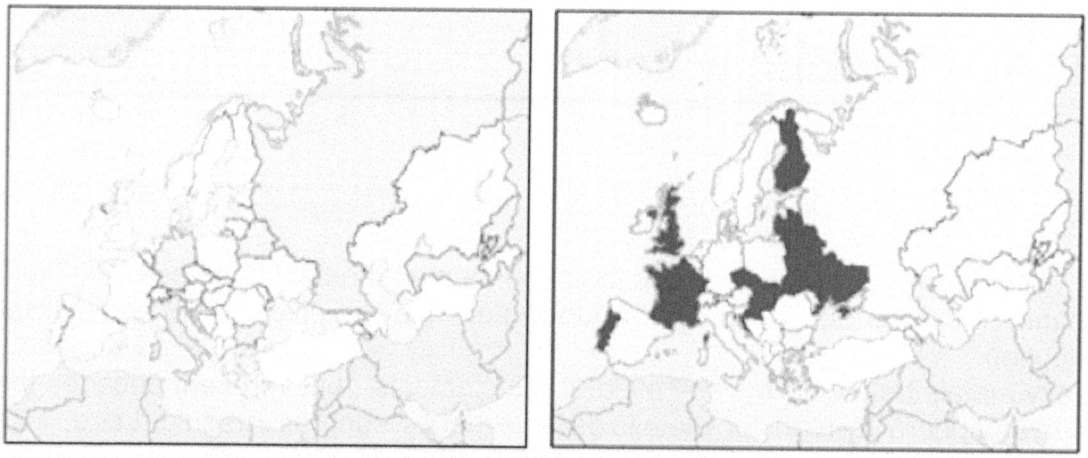

Fig. 1.1: Main drinking water problems identified by national reports (toxic: ▭ and metals: ■)

References:

- Craun G.F., McCabe L.J. (1975). Problems Associated with Metals in Drinking Water. *Journal AWWA*, **67**(11):593-99.
- Deveau M. (2010). Contribution of drinking water to dietary requirements of essential metals. *J Toxicol Environ Health A.;* **73**(2):235-41.
- Sciacca S, Oliveri Conti G. (2009). Mutagens and carcinogens in drinking water. *Mediterranean Journal Of Nutrition and Metabolism*; **2**:157-62.
- Sciacca S., Ferrante M., Oliveri Conti G. (2011). Mutagens and Carcinogens in Water Resources. Nova Science Publisher Inc. 400 Oser Avenue, Suite 1600, Hauppauge New York. ISBN: 978-1-61324-599-6 2011.
- Trent Z., Thyssen N. (2003). (WEU10) *Drinking Water Quality*. European Environmental Agency. http://www.eea.europa.eu/data-and-maps/indicators/drinking-water-quality-1/drinking-water-quality.

Chapter 2
Metals in water resources.

Margherita Ferrante, Gea Oliveri Conti.

Increased urbanization and water demand in areas of industrial activity has increased the problem of metals in groundwater sources. In small quantities, certain metals are nutritionally essential for a healthy life, but large amounts of any of them may cause acute or chronic toxicity. Heavy metals are also part of the manufacturing process of many common household items, such as pesticides, batteries, electronics, electroplated metal parts, textile dyes and steel.

Metal pollution is an important consideration for Integrated Water Resource Management (IWRM). IWRM is a process that promotes the coordinated development and management of water, land and related resources in order to maximise the resultant economic and social welfare in an equitable manner without compromising the sustainability of vital ecosystems. Freshwater systems are especially important because they represent 0.0001% of the world's water supply and are subject to increasing pressure from competing interests involved in social, economic, political and ecological activities (Gleick, 2000). Inherent in IWRM is the need to know levels and sources of contamination, that threaten the equitable apportionment of water resources between competing demands and integral aspects of water resource management. Integrated water resource management has in the past been perceived as a Government responsibility with its appointed institution as the sole stakeholder. However, what has become clear is that everyone, no matter their role in society, has a part to play in order to ensure the sustainability of our natural resources for the future.

References:

- Gleick P. (2000). The World's Water 2000–2001. Washington D.C., Island Press, 2000.
- Global Water Partnership Technical Advisory Committee - IWRM (GWPTAC). (2000). Publ. Global Water Partnership, Stockholm, Sweden.

Chapter 3
Metals And Health

Zorica Rasic-Milutinovic, Dragana Jovanovic.

Metals and metalloids can reach drinking water from various sources such as natural, industrial and agricultural use, pipe corrosion and leaching from metals in contact with water. Depending on concentration in potable water, metals and metalloids might have significant impact on overall human exposure.

Some metals are essential and some are hazardous for body functioning. Calcium, magnesium, potassium, and sodium should be present for maintenance of water body and acid-base balance, electric charge of the cell membrane, neuromuscular excitability and playing an active role in endocrine system, neuromuscular functions and in a number of enzyme activations. Iron, manganese, molybdenum, cobalt, copper, selenium, and zinc at low concentrations are needed in the human body as integral part of metalloenzymes participating in numerous biochemical and metabolic processes. In the case of high levels or lack of these essential substances adverse health effects may occur.

Drinking water might be source of chronic exposure to toxic heavy metals and metalloids that the body does not require such as arsenic, cadmium, hexavalent chromium , lead, mercury and nickel. The main route of exposure from drinking water is ingestion. The toxicity of these metals may depend on various factors e.g. type of metal, its chemical and physical form, routes of exposure, duration and dose of exposure, toxicodynamics and toxicokinetics, and also individual variability in metabolism (e.g. methylation capacity of arsenic), age, nutritional status, socio-economic status, etc. Heavy metals chronic toxicity from drinking water exposure pathway includes a wide range of adverse health effects; nearly all organs systems are involved, mostly central nervous, cardiovascular, hematopoietic, gastrointestinal and renal system. Arsenic, cadmium, hexavalent chromium and nickel have also additional confirmed carcinogenic effects and increase the risk of getting cancer.

After intestinal absorption most of heavy metals accumulate in liver, kidney, bones, soft tissue and hair. Body burden by heavy metals can be determined by measuring biomarkers of exposure mostly in blood, urine, nails and hair.

Food is a more dominant source of heavy metals exposure than drinking water. However, absorption of soluble heavy metals compounds from drinking water is higher than that from food.

If heavy metals are present above maximum contaminants levels they should be removed from drinking water by adequate treatment technique to protect human health. Instead, there are still insufficient scientific data and great uncertainty on health risk associated with an exposure to toxic heavy metals and metalloids at low-levels, that is common exposure scenario regarding drinking water exposure routes.

References:

- WHO (World Health Organization). (2011). Guidelines for Drinking-water Quality. 4th ed. WHO, Geneva, Switzerland. ISBN 978 92 4 154815 1.
- WHO (World Health Organization). (2009). Calcium and magnesium in drinking water: public health significance. Geneva, Switzerland.
- WHO (World Health Organization). (2005). Nutrients in Drinking Water. Geneva, Switzerland.
- WHO (World Health Organization). (1992). Cadmium. Environmental Health Criteria, vol. 134. Geneva, Switzerland.
- WHO (World Health Organization). (1991). Inorganic Mercury. Environmental Health Criteria, vol. 118. Geneva, Switzerland.
- WHO (World Health Organization). (1995). Lead. Environmental Health Criteria, vol. 165. Geneva, Switzerland.
- WHO (World Health Organization). (2001). Arsenic and Arsenic Compounds. Environmental Health Criteria, vol. 224. Geneva, Switzerland.
- ASTDR (Agency for Toxic Substances and Disease Registry). (2007). Toxicological Profile for Arsenic. Atlanta, GA.
- ASTDR (Agency for Toxic Substances and Disease Registry). (2008). Toxicological Profile for Cadmium. Atlanta, GA.
- ASTDR (Agency for Toxic Substances and Disease Registry). (2007). Toxicological Profile Lead. Atlanta, GA.
- ASTDR (Agency for Toxic Substances and Disease Registry). (1999). Toxicological Profile for Mercury. Atlanta, GA.
- ASTDR (Agency for Toxic Substances and Disease Registry). (2008). Toxicological Profile for Chromium. Atlanta, GA.
- ASTDR (Agency for Toxic Substances and Disease Registry). (2005). Toxicological Profile for Nickel. Atlanta, GA. Tox CAS 7440-38-2.

Chapter 4
Toxic metals

Zorica Rasic-Milutinovic, Dragana Jovanovic.

Heavy metals are commonly defined as those having a specific density of more than 5 g/cm^3 (Jarup, 2003). The Metalloid Arsenic is frequently included in this group because of its toxicity. The origin of the metals in drinking water might be from water resources (e.g. arsenic, chromium, cadmium), from the distribution system (e.g. lead, copper, cadmium) and tap (e.g. nickel, chromium). In order to protect human health from heavy metals threat, the European Union has set the maximum contaminant levels (MCL) for the most common toxic heavy metals that can be found in drinking water. In Consequence, public water supplies are regularly monitored for the following metals: arsenic (As), antimony (Sb), boron (B), cadmium (Cd), copper (Cu), chromium (Cr), lead (Pb), mercury (Hg), nickel (Ni), selenium (Se) and as indicator parameters: aluminum (Al), iron (Fe), and manganese (Mn) (Council Directive 98/83/EC), these MCL are most protective against the levels of new guideline of WHO (see Table 41.1) (WHO, 2011).

Health concerns associated with heavy metals in drinking water arise primarily from their ability to cause adverse health effects, mostly, after prolonged periods of exposure (see chapter 1).

The toxic effects of metals depend on exposure level, route of exposure, period of exposure, chemical form, bioavailability, as well as on the individual's age, nutritional and health status. Passing through human body, various factors such as absorption, distribution, metabolism and excretion influence their toxicity. Most of the heavy metals bind to sulfhydryl groups thus inhibiting enzyme activity, disrupting cellular transport and causing changes in protein functions. The toxicity of heavy metals includes the blocking of active groups of important functional molecules, e.g. enzymes, polynucleotides, transport systems for essential nutrients and ions, and substitution of essential ions from cellular sites (Kakkar and Jaffery, 2005). Health risk assesssment of heavy metals or any chemical requires an accurate exposure assessment and determination of quantitative relationship between internal dose of metal and adverse health effects and known dose-response relationship. To assess magnitude of human exposure to toxic heavy metals, we measure its internal dose in various biological materials such as blood, urine, (biomarkers) (Table 4.1).

Table 4.1. The accumulation sites in human tissues, biomarkers of exposure and chronic non-cancerogenic toxic effects of several heavy metals that can be found in drinking water.

Metal	Accumulation sites in human tissues	Typical biomarkers of exposure	Chronic non-cancerogenic toxic effects	MCL[••] in drinking water mg/l
As	Hair, nails, brain, kidney	Urine, blood, cord blood: total or iAs, MMA*, DMA**, TMA*** Hair: iAs Nail: iAs	Dermal lesion, peripheral vascular disease, cardiovascular diseases, type 2 diabetes, adverse pregnancy outcomes, respiratory diseases (asthma), adverse immune response	0.01
Cd	Bones, kidney	Urine: Cd, blood: Cd	Kidney dysfunction, Itai-itai disease (osteomalacia, osteoporosis, bone fractures)	0.005
Cr (VI)	Kidney, liver, bones	Urine: total Cr, Blood, RBCs: Cr	Hypotension, hepatic and renal failure, reproductive toxicity	0.05
Ni	Lung, liver, kidney	Urine: Ni Blood: Ni	Dermatitis (eczema), respiratory deseases, reproductive toxicity, neurotoxicity, immunotoxicity	0.02
Hg	Brain, hair, nails	Urine: Hg Blood: Hg	Tubular necrosis, proteinuria, hypoalbuminemia, neurotoxicity	0.001
Pb	Bones, teeth, hair, nails	Blood, cord blood, finger stick: Pb, Urine: Pb, ALA•	Kidney damage, cognitive impairment, anemia, reproductive toxicity, hypertension	0.01

iAs - inorganic arsenic
RBCs-red blood cells
* Methylarsonic acid
** Dimethylarsonic acid
*** Trimethylarsonic compounds
•δ-aminolevulinic acid
••According to Council Directive 98/83/EC

Hereafter, chronic mostly non-carcinogenic toxic effects of several metals including arsenic, cadmium chromium, nickel, mercury and lead will be presented as drinking water is a well recognized pathway of exposure to these metals. For more specific health advice the reader should consult a suitably qualified medical practitioner.

The current knowledge in the field of metals-biochemistry of oxidative stress indicates that metal-induced and metal-enhanced formation of free radicals and other reactive species can be regarded as a common factor in determining metal-induced toxicity together with their carcinogenicity (Jomova & Valko, 2011). Recently, more attention and concern has been given to metal compounds that have toxic effects at low levels of exposure than those that produce overt clinical and pathological signs and symptoms (Kalia & Flora, 2005). The role of metals as endocrine disruptors have recently been studied, too. This means that some metals can interfere with hormone biosynthesis, secretion and metabolism that may lead to adverse health outcomes such as reproductive disorders, thyroid and neuro-developmental outcomes and endocrine-related cancers. Cadmium and lead have been the most studied metals in relation to altered hormone levels. Cadmium is recognized as an endocrine disruptor but the mechanisms involved are not well understood (Meeker et al., 2010). Positive associations between low-level cadmium exposure and increased FSH and testosterone in men, and estradiol and FSH in postmenopausal women have been recently reported (Menke et al., 2008; Nagata et al., 2005; Zeng et al., 2004). The existing evidences of a relationship between exposure to metals and hormone levels are inconsistent, and there is lack of studies on hormone alterations related to exposure to metals among the general population (Meeker et al., 2010).

References:

- Jarup L. (2003). Hazards of heavy metal contamination. *British Medical Bulletin*; **68**: 167–82.
- Jomova K., Valko M. (2011). Advances in metal-induced oxidative stress and human disease. *Toxicol*; **283**:65–87.
- Kalia K., Flora S.J. (2005). Strategies for safe and effective therapeutic measures for chronic arsenic and lead poisoning. *J Occup Health*; **47**:1–21.
- Meeker J.D., Rossano M.G., Protas B., Padmanabhan V., Diamond M.P., Puscheck E., Daly D., Paneth N., Wirth J.J. (2010). Environmental exposure to metals and male reproductive hormones: Circulating testosterone is inversely associated with blood molybdenum. *Fertil Steril*; **93**(1):130.
- Menke A., Guallar E., Shiels M.S., Rohrmann S., Basaria S., Rifai N., Nelson W.G., Platz E.A. (2008). The association of urinary cadmium with sex steroid hormone concentrations in a general population sample of US adult men. *BMC Public Health*;8:72.
- Nagata C., Nagao Y., Shibuya C., Kashiki Y., Shimizu H. (2005). Urinary cadmium and serum levels of estrogens and androgens in postmenopausal Japanese women. *Cancer Epidemiol Biomarkers Prev*; **14**(3):705–8.
- Official Journal of the European Communities L 330/33. Council Directive 98/83/EC.

- Kakkar P., Jaffery F.N. (2005). Biological markers for metal toxicity. *Environmental Toxicology and Pharmacology*; **19**:335–49.
- Zeng X., Jin T., Buchet J.P., Jiang X., Kong Q., Ye T., Bernard A., Nordberg G.F. (2004). Impact of cadmium exposure on male sex hormones: a population-based study in China. *Environ Res*; **96**(3):338–44.

Chapter 5
Mutagenic and genotoxic metals

Zorica Rasic-Milutinovic, Dragana Jovanovic.

A chemical is considered to be mutagenic if it is capable of inducing heritable changes (mutations) in the genotype of a cell as a consequence of alterations or loss of genes, chromosomes or parts of chromosomes. Genotoxicity is a broader term that refers to the ability to interact with DNA and/or the cellular structures that regulate the fidelity of the genome, such as the spindle apparatus and topoisomerase enzymes (Worth & Balls, 2002). Most human carcinogens are genotoxic but not all genotoxic agents have been shown to be carcinogenic for humans (Albertini et al., 2000). Some epidemiological and experimental studies indicate that toxic heavy metals may act at different stages in the carcinogenic process and that several different mechanisms may be involved such as formation of DNA adducts and DNA strand breaks, mutations in genes, chromosomal aberrations, aneuploidy, and changes in DNA methylation patterns (Caldwell, 2012). Genotoxicity of heavy metals or any chemical is tested in experimental conditions through in vitro and in vivo experimental models and study results may vary with type of metals and compounds tested, specificity of test systems used and experimental conditions (EPA, 2012). According to US National Toxicology Program (NTP) genotoxic chemicals are likely to exhibit trans-species carcinogenicity, often in both sexes, at intermediate dose levels, and this will not necessarily be restricted to one target tissue (Ashby & Tennant, 1991; Worth & Balls, 2002). For genotoxic carcinogens it is considered that even very low levels of exposure may increase the risk of adverse outcomes; no exposure threshold is assumed in the cancer risk assessment process.
Hereafter, mutagenic and genotoxic effects of several metals including arsenic, cadmium chromium, nickel, mercury and lead will be shortly presented in dedicated sections.

References:

- Albertini R.J., Anderson D., Douglas G.R., Hagmar L., Hemminki K., Merlo F., Natarajan A.T., Norppa H., Shuker D.E., Tice R., Waters M.D., Aitio A. (2000). IPSC Guideline for monitoring of genotoxic effects of carcinogens in humans. *Mutat Res*; **463**(2): 111-72.
- Ashby J., Tennant R.W. (1991). Definitive relationships among chemical structure, carcinogenicity and mutagenicity of 301 chemicals tested by the US NTP. *Mutation Research*; **257**, 229-306.
- Caldwell J.C. (2012). DEHP: Genotoxicity and potential carcinogenic mechanisms - A review, *Mutat. Res.: Rev. Mutat. Res,* http://dx.doi.org/10.1016/j.mrrev.2012.03.001. (accessed 1 March 2012).

- U.S.EPA. (2005a). Guidelines for carcinogen risk assessment. Risk Assessment Forum, Washington, DC; EPA/630/P-03/001B. Available online at http://www.epa.gov/iris/backgrd.html (accessed April 10, 2012).
- Worth A.P., Balls M. ed. (2002). Genotoxicity and Carcinogenicity. In: Alternative (non-animal) Methods for Chemicals Testing: Current Status and Future Prospects. *ATLA* **30**(1)125:83-93.

Chapter 6
Carcinogenic metals

Zorica Rasic-Milutinovic, Dragana Jovanovic.

Defining a mechanism of metal carcinogenesis is critical for human health risk assessment. For many metals, aberrant cell proliferation, including alterations in apoptosis is an attractive aspect of a hypothetical mechanism of cancer induction. Apoptosis is considered as an ongoing normal event in the control of cell populations. Apoptosis essentially occurs when cellular damage, including damage to genetic material, has exceeded the capacity for repair. Environmental metals can impair apoptosis thus facilitating aberrant cell accumulation that may be a critical step in the pathogenesis of malignancy or autoimmunity (Vir & Rana, 2008).

Heavy metals are reported to be tumor promotors (Rhee et al., 2000; Wu et al., 1999). They are thought to promote the *mutagenic effects* of DNA damaging agents, while alone may not themselves be mutagenic (Goyer, 1996: Maier et al., 2002). Recently, Hirata et al. (Hirata et al., 2010) have shown that the effects of heavy metals, As^{3+}, Cr^{6+}, Pb^{2+} and Cd^{2+} on DNA annealing and unwinding activities are mediated, at least in substantial part, through actions of the mono-ubiquitinated annexin A1 homodimer.

Heavy metal induced *carcinogenesis* is well documented by epidemiological studies, and several diverse mechanisms of cancer induction are postulated, depending on heavy metals and exposed tissues (Galanis et al., 2009). The tumor promoting action of heavy metals is attributed to enhanced signals of growth factors for proliferation (Wu et al., 1999), and perturbations in nucleosome structure by heavy metals are implicated for genetic and epigenetic alterations in cancers (Mohideen et al., 2010). It has been emphasized that exposure to heavy metals causes elevated leves of oxidative stress (Kasprzak, 1995; Calsou et al., 1996; Galanis et al., 2009; Durackova, 2010). Consistent with the data of Hirata A. et al., annexin A1 is proposed to be involved in intracellular signaling during stress that results in profound changes in Ca^{2+} and pH homeostasis (Monastyrskaya et al., 2009). It has been well documented, that in various cancer tissues free radical-mediated DNA damage has occurred (Marnett, 2000).

DNA analysis from colon and rectal biopsies revealed that iron-induced oxidative stress may be the key determinant of human colorectal cancer (Nelson, 1992; Skrzydlewska et al., 2005). Since copper is known to promote oxidative stress and inflammation, it is likely that under nonphysiological conditions of increased copper levels, it could play a role in the development of various cancers (breast, cervical, ovarian, lung, prostate, stomach cancer and leukemia) (Gupte & Mumper, 2009). It has been also shown that the Cu: (Zn, Se, Fe) ratios are frequently higher in cancer patients compared to normal subjects (Gupte and Mumper, 2009).

Chromium(VI) at high doses is considered to be a great health risk (Keegan et al., 2008). An increased rate in stomach tumours was observed in humans and animals exposed to Cr(VI) in drinking water. Recent studies using cells cultures revealed a much greater potential for Cr(VI) to cause chromosomal damage and mutations (Reynolds et al., 2007).

Intake of cobalt significantly suppresses free radical formation, oxidation of lipids and proteins. In addition, the GSH/GSSH ratio was similar to that of control cells activated by heme oxygenase 1 (Fenoglio et al., 2008). These results look promising in view of the prospective pharmacological benefits of cobalt in preventing hypoxia-induced oxidative stress.

Displacement of copper and iron by cadmium can explain the enhanced cadmium-induced toxicity, because copper, displaced from its binding site, is able to catalyze breakdown of hydrogen peroxide via the Fenton reaction (Flora, 2009). Cadmium is a potent human carcinogen causing preferentially prostate, lung, kidney and pancreas cancers. However, metal induced formation of free radicals has most significantly been evidenced for iron and copper then for chromium and partly for cobalt.

Oxidative stress has been linked with the development of arsenic related diseases including cancers (Waalkes et al., 2004). Chronic exposure to inorganic arsenic from contaminated water is responsible for various adverse health effects such as developing tumours of lung, skin, liver, bladder and kidney.

Zinc is a redox inert metal and does not participate in oxidation-reduction reactions. However, zinc deficiency has been associated with increased levels of oxidative damage including increased lipid, protein and DNA oxidation, as well as activation of growth factors and antiapoptotic molecules resulting in cell proliferation (cancer) (Prasad, 2009).

Nickel compounds were shown to act synergistically with many mutagenic carcinogens in enhancing cell transformation both in vitro and in vivo (Ishimatsu et al, 1995). Many studies have demonstrated that some heavy metals (Ni, Cr and Cd) are complex carcinogens, and the mechanisms underlying these metals carcinogenesis are multifactorial. The major mechanisms of Ni carcinogenesis include aberrant gene expression, inhibition of DNA methylation, inhibition of DNA damage repair and apoptosis, and induction of oxidative stress (Hu et al, 2004; Schwerdtle et al. 2002). IARC classified metallic nickel in group 2B (possibly carcinogenic to humans) and nickel compounds in group 1 (carcinogenic to humans).

IARC classified the carcinogenicity rating of beryllium as a Group 1 substance as there was sufficient evidence for carcinogenicity in humans (IARC). Only a small number of studies have reported on the mutagenic potential of beryllium compounds in mammalian cells (Miyaki et al.; Hsie et al.; 1978).

References:

- Calsou P., Frit P., Bozzato C., Salles B. (1996). Negative interference of metal (II) ions with nucleotide excision repair in human cell-free extracts. *Carcinogenesis*; **17**(12), 2779–82.
- Durackov Z. (2010). Some current insights into oxidative stress. *Physiol. Res.*; **59**: 459–69.
- Fenoglio I., Corazzari I., Francia C., Bodoardo S., Fubini B. (2008). The oxidation of glutathione by cobalt/tungsten carbide contributes to hard metal-induced oxidative stress. *Free Radic. Res.*; **42**:437–45.

- Flora S.J. (2009). Structural, chemical and biological aspects of antioxidants for strategies against metal and metalloid exposure. *Oxid. Med. Cell. Longev.*; **2**:191–206.
- Galanis A., Karapetsas A., Sadaltzopoulos R. (2009). Metal-induced carcinogenesis, oxidative stress and hypoxia signaling. *Mutation Res.* **674**:31–5.
- Goyer R.A. (1996). Toxic effects of metals. In: Klassen, C.D. (Ed.), Casarett and Doull's toxicology: the basic science of poisons. McGraw Hill, New York, pp. 691–736.
- Gupte A., Mumper R.J. (2009). Elevated copper and oxidative stress in cancer cells as a target for cancer treatment. *Cancer Treat. Rev.*: **35**:32–46.
- Hirata A., Corcoran G., Fusao H. (2010). Carcinogenic heavy metals replace Ca^{2+} for DNA binding and annealing activities of mono-ubiquitinated annexin A1 homodimer. *Toxicology and Applied Pharmacology*; **248**:45-51.
- Hsie A.W. (1978). Quantitative mammalian cell genetic toxicology. *Envrion. Sci. Res.* **15**:291–315.
- Hu W., Feng Z., Tang M.S. (2004). Nickel (II) enhances benzo[a]pyrene diol epoxideinduced mutagenesis through inhibition of nucleotide excision repair in human cells: a possible mechanism for nickel (II)-induced carcinogenesis. *Carcinogenesis*; **25**:455–62.
- Kasprzak K.S. (1995). Possible role of oxidative damage in metal-induced carcinogenesis. *Cancer Invest.* **13**(4):411–30.
- Keegan G.M., Learmonth I.D., Case C.P. (2008). Asystematic comparison of the actual, potential, and theoretical health effects of cobalt and chromium exposures from industry and surgical implants. *Crit. Rev. Toxicol*; **38**:645–74.
- IARC (International Agency for Research on Cancer). IARC Monographs on the Evaluation of Carcinogenic Risks to Humans, vol. 58, Beryllium, Cadmium, Mercury and Exposures in the Glass Manufacturing Industry, IARC, Lyon, 1993, pp. 41–117.
- Ishimatsu S., Kawamoto T., Matsuno K., Kodama Y. (1995). Distribution of various nickel compounds in rat organs after oral administration. *Biol. Trace Elem. Res.* **49**:43–52.
- Maier A., Schumann B.L., Chang X., Talaska G., Puga A. (2002). Arsenic co-exposure potentiates benzo[a]pyrene genotoxicity. *Mutat. Res.* **517**(1–2):101–11.
- Marnett L.J. (2000). Oxyradicals and DNA damage. *Carcinogenesis*. **21**:361–70.
- Miyaki M., Akamatsu N., Ono T., Koyama H. (1979). Mutagenicity of metal cations in cultured cells from Chinese hamster. *Mutat. Res.* **68**(3):259–63.
- Mohideen K., Muhammad R., Davey C.A. (2010). Perturbations in nucleosome structure from heavy metal association. *Nucleic Acids Res.* **38**:6301-11.
- Monastyrskaya M., Bbiychuck E.B., Draeger A. (2009). The annexins: spatial and temporal coordination of signaling events during cellular stress. *Cell. Mol. Life Sci.* **66**:2623–42.
- Nelson R.L. (1992). Dietary iron and colorectal cancer risk. *Free Radic. Biol. Med.*; **12**:161–68.

- Prasad A.S. (2009). Zinc: role in immunity, oxidative stress and chronic inflammation. *Curr. Opin. Clin. Nutr. Metab. Care*; **12**:646–65.
- Reynolds M., Stoddard L., Bespalov I., Zhitkovich A. (2007). Ascorbate acts as a highly potent inducer of chromate mutagenesis and clastogenesis: linkage to DNA breaks in G2 phase by mismatch repair. *Nucleic Acids Res*; **35**:465–76.
- Rhee H.J., Kim G.Y., Huk J.W., Kim S.W., Na D.S. (2000). Annexin I is a stress protein induced by heat, oxidative stress and a sulfhydryl-reactive agent. *Eur. J. Biochem*. **267**(11):3220–25.
- Schwerdtle T., Seidel A., Hartwig A. (2002). Effect of soluble and particulate nickel compounds on the formation and repair of stable benzo[a]pyrene DNA adducts in human lung cells. *Carcinogenesis*; **23**:47–53.
- Skrzydlewska E., Sulkowski S., Koda M., Zalewski B., Kanczuga-Koda L., Sulkowska M. (2005). Lipid peroxidation and antioxidant status in colorectal cancer. *World J. Gastroenterol*; **11**:403–06.
- Waalkes M.P., Liu J., Ward J.M., Diwan L.A. (2004). Mechanisms underlying arsenic carcinogenesis: hypersensitivity of mice exposed to inorganic arsenic during gestation. *Toxicology*; **198**:31–38.
- Wu W., Graves L.M., Jaspers I., Devlin R.B., Reed W., Samet J.M. (1999). Activation of the EGF receptor signaling pathway in human airway epithelial cells exposed to metals. *Am. J. Physiol.* 277(5 pt 1), L924–31.
- Vir S., Rana S. (2008). Metals and apoptosis: Recent developments. *J Trace Elem Med Biol*; **22**:262-84.

Chapter 7
Aluminium (Al)

Adriana Floridia, Sanjay Mishra.

7.1 ENVIRONMENTAL EFFECT

Aluminium, atomic number 13, is ubiquitous; it is the most abundant metallic element and constitutes about 8% of Earth's crust (WHO, 2011). Aluminium metal is light in weight and silvery-white in appearance. It is naturally released to the environment from the weathering of rocks and volcanic activity but also by human activities. Aluminium is a very reactive element and it is never found as the free metal in nature as it is combined with other elements, most commonly with oxygen, silicon, and fluorine. These chemical compounds are commonly found in soil, minerals (e.g., sapphires, rubies, turquoise), rocks (especially igneous rocks), and clays (ATDSR, 2008).

Aluminium enters the atmosphere as a major constituent of atmospheric particulates originating from natural soil erosion, mining or agricultural activities, volcanic eruptions or coal combustion (WHO, 2011).

It is used for making beverage cans, pots and pans, airplanes, siding and roofing, and foil. Powdered aluminium is often used in explosives and fireworks. Powdered aluminium metal is often used in explosives and fireworks. Aluminium compounds are used in many diverse and important industrial applications such as alums (aluminium sulfate),water-treatment (as coagulants to reduce organic matter, colour, turbidity and microorganism levels) and alumina in abrasives and furnace linings (WHO, 2011). Aluminium is found in over-the-counter medicinals, such as antacids and buffered aspirin, it is used as a food additive, and is found in a number of topically applied consumer products such as antiperspirants, and first aid antibiotic and antiseptics. The concentration of aluminium in food and beverages varies widely, depending upon the food product, the type of processing used, and the geographical areas in which food crops are grown (ATDSR, 2008). Aluminium cannot be destroyed in the environment. It can only change its form or be attached or separated from particles. Aluminium particles in air settle to the ground or are washed out of the air by rain. However, very small aluminium particles can stay in the air for many days. Most aluminium-containing compounds do not dissolve to a large extent in water unless the water is acidic or very alkaline (ATSDR, 2008).

The general population is primarily exposed to aluminium through the consumption of food items, although minor exposures may occur through ingestion of aluminium in drinking water and inhalation of ambient air. Aluminium intake from foods, particularly those containing aluminium compounds used as food additives, represents the major route of aluminium exposure for the general public, excluding persons who regularly ingest aluminium-containing antacids and buffered analgesics (WHO, 1997). Unprocessed foods like fresh fruits, vegetables, and meat contain very little aluminium. Aluminium compounds may be added during processing of foods, such as: flour, baking powder, coloring agents, anticaking agents. The concentration of aluminium in natural waters (e.g., ponds, lakes,streams) is generally below 0.1 milligrams per liter (mg/L). People generally consume little aluminium from drinking water. Aluminium levels in drinking-water vary according to the levels found in the source water and whether aluminium coagulants are used during treatment to become drinking water (WHO, 2011). Water is sometimes treated with aluminium salts while it is processed but even then, aluminium levels generally do not exceed 0.1 mg/L (ATDSR, 2008). The Environmental Protection Agency (EPA) has recommended a Secondary Maximum Contaminant Level (SMCL) of 0.05–0.2 mg/L for aluminium in drinking water. The SMCL is not based on levels that will affect humans or animals. It is based on taste, smell, or color (ATDSR, 2008).

7.2 EFFECT ON HUMAN HEALTH

There is little indication that aluminium is acutely toxic by oral exposure despite its widespread occurrence in foods, drinking-water and many antacid preparations (WHO, 1997). Aluminium has been associated with several neurodegenerative diseases, such as dialysis encephalopathy, amyotrophic lateral sclerosis and Parkinsonism dementia, and in particular, Alzheimer's disease (Kawahara, 2005; Becaria et al., 2002). Aluminium was identified, along with other elements, in the amyloid plaques that are one of the diagnostic lesions in the brain for Alzheimer disease. Numerous epidemiological studies have been carried out to try to determine the validity of this hypothesis, most of these have focused on aluminium in drinking-water but the results are highly contradictory (Bondy, 2010; Flaten, 2001; ATDSR, 2008). It is possible that the causal role of aluminium in AD may have to be reconsidered (Kawahara, 2005; Andrasi et al., 2005). Aluminiumhas been implicated in the formation of neurofibrillary tangles (Walton, 2006) but it is unclear which came first, the tangles or the Aluminium. The general medical opinion is that the association is an epi-phenomenon without causative consequences Although aluminium-containing 'over the counter' oral products are considered safe in healthy individuals at recommended doses, some adverse effects have been observed following long-term use in some individuals. Some people who have kidney disease store a lot of aluminium in their bodies. Sometimes, these people developed bone or brain diseases that can be caused by the excess aluminium. Brain and bone disease caused by high levels of aluminium in the body have been seen in children with kidney disease. Bone disease has also been seen in children taking some medicines containing aluminium. In these children, the bone damage is caused by aluminium in the stomach preventing the absorption of phosphate, a chemical compound required for healthy bones (ATSDR, 2008).Workers who breathe large amounts of aluminium dusts can have lung problems, such as coughing or changes that show up in chest X-rays. Some workers who breathe aluminium-containing dusts or aluminium fumes have decreased performance in some tests that measure functions of the nervous system. (ATDSR, 2008, WHO, 2011). The Joint FAO/WHO Expert Committee on Food Additives (JECFA) in 2007, developed a provisional tolerable weekly intake (PTWI) for aluminium from all sources of 1 mg/kg of body weight (FAO/WHO, 2007).

References:

- ATSDR. (2008). Toxicological profile for aluminum. Atlanta, GA, United States Department of Health and Human Services, Public Health Service, Agency for Toxic Substances and Disease Registry.
- Becaria A., Campbell A., Bondy S.C. (2002). Aluminum as a toxicant. *Toxicol Ind Health*; **18**(7):309-20.
- Bondy S.C. (2010). The neurotoxicity of environmental aluminum is still an issue. *Neurotoxicology*; **31**(5):575-81.
- FAO/WHO. (2007). In: Aluminum (from all sources, including food additives). In: Evaluation of certain food additives and contaminants. Sixty-seventh report of the Joint FAO/WHO Expert Committee on Food Additives. Geneva, World Health Organization, pp. 33–44.
- Flaten T.P. (2001). Aluminium as a risk factor in Alzheimer's disease, with emphasis on drinking water. *Brain Res Bull*, 15,**55**(2):187-96.
- WHO (World Health Organization). (1997). Aluminium. Geneva, International Programme on Chemical Safety (Environmental Health Criteria 194), Geneva.
- WHO (World Health Organization). (2011). Guidelines for Drinking-water Quality. 4th ed. WHO, Geneva, Switzerland. ISBN 978 92 4 154815 1.
- Kawahara L. (2005). Effects of aluminum on the nervous system and its possible link with neurodegenerative diseases. *M. J Alzheimers Dis*. **8**(2):171-82.

Chapter 8
Antimony (Sb)

Michael R. Moore

8.1 ENVIRONMENTAL EFFECT

Antimony, atomic number 51, is an extremely brittle metalloid of a flaky, crystalline texture. It is bluish white and has a metallic lustre. It is not acted on by air at room temperature, but burns brilliantly when heated with the formation of white fumes. It is a poor conductor of heat and electricity. The largest applications for metallic antimony are as alloying material for lead and tin and for lead antimony plates in lead-acid batteries. Alloying lead and tin with antimony improves the properties of the alloys which are used in solders, bullets and plain bearings. Antimony compounds are prominent additives for chlorine- and bromine-containing fire retardants found in many commercial and domestic products. An emerging application is the use of antimony in microelectronics.

Antimony and its compounds are toxic. It is found mostly with other minerals and in stibnite.

8.2 EFFECT ON HUMAN HEALTH

Harmful effects of antimony upon body tissues and functions can occur following inhalation or ingestioin of certain compounds of antimony. Such poisoning resembles arsenic poisoning. Improvements in working conditions have remarkably decreased the incidence of antimony toxicity in the workplace.

Antimony poisoning has resulted from drinking acidic fruit juices containing antimony oxide dissolved from the glaze of enamelware containers. Toxicity can also result from repeated exposure to antimony in medications, such as tartar emetic (antimony and potassium tartrate), used to induce vomiting and in treatment of helminthic and fungal infestations (ATSDR, 1992) .

Antimony had a reputation of being a universal panacea of all kinds of diseases in the middle ages. In the past antimony compounds have been used for the treatment of two parasitic diseases schistosomiasis and leishmaniasis.

The major toxic side-effects of antimonials as a result of therapy are cardiotoxicity (~9% of patients) and pancreatitis, which is seen commonly in HIV and visceral leishmaniasis co-infections.

Antimony and many of its compounds are toxic, and the effects of antimony poisoning are similar to arsenic poisoning but the toxicity of antimony is lower than that of arsenic. This might be caused by the significant differences of uptake, metabolism and excretion between arsenic and antimony. Antimony toxicity occurs either due to occupational exposure or during therapy. Occupational exposure may cause respiratory irritation, pneumoconiosis, antimony spots on the skin and gastrointestinal symptoms. In addition antimony trioxide is possibly carcinogenic to humans The uptake of antimony(III) or antimony(V) in the gastrointestinal tract is at most 20%. The most serious effect of acute antimony poisoning is cardiotoxicity and the resulted myocarditis. Inhalation of antimony dust is harmful and in certain cases may be fatal; in small doses, antimony causes headaches, dizziness, and depression. Larger doses such as prolonged skin contact may cause dermatitis, or damage the kidneys and the liver, causing violent and frequent vomiting, leading to death in a few days.

Antimony leaches from polyethylene terephthalate (PET) bottles into liquids. While levels observed for bottled water are below drinking water guidelines, fruit_juice concentrates produced in the UK were found to contain up to 44.7 µg/L of antimony, well above the EU limits for tap water of 5 µg/L(Hansen et al., 2010).

The guideline value of WHO for Antimony in drinking-water is 0.02 mg/l (WHO, 2011).

References:

- ATSDR, (1992).*www.atsdr.cdc.gov/substances/toxsubstance.asp?toxid=58*
- Hansen C., Tsirigotaki A., Bak S.A., Pergantis S.A., Stürup S., Gammelgaard B., Hansen H. (2010). Elevated antimony concentrations in commercial juices. *Journal of Environmental Monitoring,* **12**:822–4.
- Sundar S., Jaya C., Int J. *(2010).* Antimony Toxicity*. J. Environ. Res. Public Health*, **7**: 4267-4277.
- WHO (World Health Organization). (2011). Guidelines for Drinking-water Quality. 4th ed. WHO, Geneva, Switzerland. ISBN 978 92 4 154815 1.

Chapter 9
Arsenic (As)

Caterina Ledda, Zorica Rasic-Milutinovic, Dragana Jovanovic.

9.1 ENVIRONMENTAL EFFECT

Arsenic, atomic number 33, is a widely distributed metalloid in the environment. It can be found naturally world-wide in small concentrations.

It occurs in soil and minerals and it may enter air, water and land through wind-blown dust and water run-off. Arsenic in the atmosphere comes from various natural sources as volcanoes, microorganisms release (volatile methylarsines) and anthropogenic sources such as burning of fossil fuels, uncontrolled release of industrial wastes and release of pesticides in agriculture, but also it comes as a by-product of refining the ores of other metals, such as copper and lead. The other key issue in recent years beside As-contaminated groundwater is that relatively high inorganic As can be found in many brands of rice and rice products.

Arsenic is persistent in environment, plants absorb arsenic fairly easily, so arsenic may bioaccumulate in food seafood and in vegetables.

The WHO Provisional guideline value of Arsenic in drinking water is 0.01 mg/l (WHO, 2011). Arsenic is a enhancer of alteration of genetic materials of fish and of all animals that eat polluted fish (ATSDR, 2007). Arsenic can reach drinking water through its occurrence, primarily, in groundwater sources as a result of geological and hydrogeological processes. Generally, inorganic As species, such as, As(III) and As(V) are more toxic, than organic species of As such as monomethylarsonic acid (MMA^{V}) and diemethylarsinic acid (DMA^{V}) (Ng, 2005).

9.2 EFFECT ON HUMAN HEALTH

Arsenic is toxic to both plants and animals and inorganic arsenicals are proven carcinogens in humans (Ng, 2005; Murcott 2012). The toxicity of arsenic to human health ranges from skin lesions to cancer of the brain, liver, kidney, and stomach (Smith et al., 1992). A wide range of arsenic toxicity has been determined that depends on arsenic speciation. Generally inorganic arsenic species are more toxic than organic forms to living organisms, including humans and other animals (Goessler & Kuehnett, 2002; Meharg and Hartley-Whitaker, 2002 ; Ng, 2005). The oral LD_{50} for inorganic arsenic ranges from 15–293 mg (As) kg^{-1} and 11–150 mg (As) kg^{-1} bodyweight in rats and other laboratory animals respectively (Done and Peart, 1971; Ng, 2005).

Exposure to arsenic trioxide by ingestion of 70–80 mg has been reported to be fatal for humans (Vallee et al., 1960). Arsenite (iAs^{III}) is usually more toxic than arsenate (iAs^{V}). Recent studies found that Monomethylarsenous acid (MMA^{III}) and Dimethylarsenous acid

(DMA^{III}) are more acutely toxic and more genotoxic than their parent compounds (Mass et al., 2001; Petrick et al., 1993; Petrick et al., 2000). These trivalent arsenicals are more toxic than iAs^{V}, MMA^{V} and DMA^{V} in vitro (Styblo et al., 2000; Mass et al., 2001). This may be related to more efficient uptake of trivalent methylated arsenicals than of pentavelent arsenicals by microvessel endothelial cells and CHO (Chinese Hamster Ovary) cells (Hirano et al., 2003; Dopp et al., 2004).

Recently, LC_{50} values were calculated as 571, 843, 5.49, and 2.16 μM for iAs^{V}, DMA^{V}, iAs^{III}, and DMA^{III}, respectively, for human cells (Naranmandura et al., 2007). This study also showed that dimethylmonothioarsenic ($DMMTA^{V}$) is much more toxic than other pentavalent nonthiolated arsenicals (Naranmandura et al., 2007).

The toxicity of trivalent arsenic is related to its high affinity for the sulfhydryl groups of biomolecules such as glutathione (GSH) and lipoic acid and the cysteinyl residues of many enzymes (Aposhian & Aposhian, 2006). The formation of As(III)–sulfur bonds results in various harmful effects by inhibiting the activities of enzymes such as glutathione reductase, glutathione peroxidases, thioredoxin reductase, and thioredoxin peroxidase (Schuliga et al., 2002; Wang et al., 1997; Lin et al., 2001; Chang et al., 2003). An example of As^{III}–S bond formation is the 1:3 complex of As with Cys-containing tripeptide GSH, which has an unusually high stability constant (Rey et al., 2004). Such As^{III}-GSH conjugates have been detected in the bile of rats (Suzuki et al., 2001). Stable arsenic complexes with the common reductant, dithiothreitol, and other dithiols, are known to exist (Zahler and Cleland, 1968; Kolozsi et al., 2008). The higher toxicity of MMA^{III} than iAs^{III} may be caused by a higher affinity of MMA^{III} for thiol ligands in biological binding sites than As^{III}–thiolate complexes (Spuches et al., 2005). DMA^{III} also forms complexes with sulfur-rich proteins (Shiobara et al., 2001; Naranmandura et al., 2006).

It is generally accepted that pentavalent arsenicals do not directly bind to sulfhydryl groups to cause toxic effects (Suzuki et al., 2008). However, a recent study reported that sulfide-activated pentavalent arsenic could bind to the sulfhydryl group of GSH (Raab et al., 2007). An exposure of DMA^{V} to cabbage (Brassica oleracea) gave dimethylmonothioarsinic acid-GSH conjugate ($DMMTA^{V}$-GSH). $DMMTA^{V}$ was found in the urine of arsenic-exposed humans and animals (Raml et al., 2007; Naranmandura et al., 2007) and showed distinct behavior and toxicity in vivo and in vitro relative to those of the corresponding oxo acids (Suzuki et al., 2007; Naranmandura et al., 2007; Raml et al., 2007). Interestingly, $DMMTA^{V}$ demonstrated a significantly higher cytotoxicity than nonthiolated DMA^{V} (Raab et al., 2007; Raml et al., 2007). Moreover, the toxicity of $DMMTA^{V}$ is comparable to that of trivalent arsenicals. The toxicity of $DMMTA^{V}$ may be caused by the production of reactive oxygen species (ROS) during its exposure, which may cause mutagenesis and DNA damage, initiating cancer (Kitchin, 2001). A mechanism has been proposed to suggest the production of ROS through the redox equilibrium between DMA^{V} and DMA^{III} in the presence of GSH (Naranmandura et al., 2007; Suzuki et al., 2008).

Using lower bound and upper bound As concentrations based on dietary and drinking-water samples collected from 19 European countries, the CONTAM Panel (The European Food Safety Authority Panel on Contaminants in the Food Chain) estimated that EU-wide inorganic arsenic exposures range from 0.13 to 0.56 μg/kg body weight (b.w.) per day for average consumers, and 0.37 to 1.22 μg/kg b.w for 95th percentile consumers (EFSA, 2009). [WHO/FAO JECFA has a more up to date review in 2010/2011 at their 72nd meeting; and reported a $BMDL_{0.5}$ of 3 μg/kg b.w. per day based on arsenic-induced lung cancer.]

(see: The references to key epi studies by Chen et al 2010 (a; b) can be found in the JECFA 2011 technical report (the 72nd meeting)
The effects of arsenic are varied as already abovementioned. Specifically among the non-carcinogenic effects there are: dermal lesions such as hyperpigmentation, hypopigmentation, keratosis mainly on the palms and soles and are sensitive indicators of chronic inorganic arsenic ingestion and are often used as diagnostic criteria for arsenicosis; peripheral vascular disease; cardiovascular diseases; type 2 diabetes; adverse pregnancy outcomes; respiratory diseases (asthma); adverse immune response (Kozul et al., 2009; Chen and Ahsan, 2004, Wu, 1989). Numerous epidemiological studies have been conducted with strong evidence of causal association between As exposure in drinking water at concentrations above 100 μg/l and skin and internal cancer, particularly cancers of the urinary bladder, lung and kidney in adult (Wu et al., 1989; Smith et al., 1998; Hopenhayn-Rich et al., 1998; Ferreccio et al., 2000 (should cite more recent studies done in Taiwan and Bangladesh; particularly studies reported in 2009-2010 that are regarded as more robust epi studies). The associations between aforementioned disorders and drinking water arsenic seem to be well-resolved for relatively high As exposures, i.e., > 100 μg/L, however, there are many uncertainties concerning the health effects of low doses of arsenic and its possible dose-response relationships.
Genetic toxicity studies have shown that arsenic causes gene mutations, gene amplification and mitotic arrest (reaction with tubulin) as aforementioned. Arsenic induced also chromosomal aberrations, including micronuclei and aneuploidy and sister-chromatid exchanges (SCEs), enhances oxidative stress and influences the production of nitrogen monoxide. Methylated and dimethylated forms, also exibit genotoxicity at higher exposure levels. Arsenic does not cause direct DNA damage, but inhibits DNA synthesis and repair and also affects DNA methylation in tumor suppressor genes (Brown, 2008).

References:

- Aposhian H.V., Aposhian M.M. (2006). Arsenic toxicology: Five questions. *Chemical Research in Toxicology*, **19**(1):1-15.
- ATSDR. (2007). Arsenic. Division of Toxicology and Environmental Medicine ToxFAQsTM, Atlanta, GA, United States Department of Health and Human Services, Public Health Service, Agency for Toxic Substances and Disease Registry.
- Brown J.P. (2008). Risk assessment for arsenic in drinking water. In: Howd R.A. and Fan A.M. (Eds), Risk assessment for chemicals in drinking water. New York: Wiley & Sons; pp.228-30.
- Chang K.N., Lee T.C., Tam M.F., Chen Y.C., Lee L.W., Lee S.Y., Lin P.J., Huang R.N. (2003). Identification of galectin I and thioredoxin peroxidase II as two arsenic-binding proteins in Chinese hamster ovary cells. *Biochemical Journal*, **371**(2):495-503.
- Chen Y., Ahsan H. (2004). Cancer burden from arsenic in drinking water in Bangladesh. *American Journal of Public Health*; **94**:741-4.
- Done A.K., Peart A.J. (1971). Acute toxicities of arsenical herbicides. *Clinical toxicology*, 4(3): 343-55.
- Dopp E., Hartmann L.M., Florea A.M., Von Recklinghausen U., Pieper R.,

Shokouhi B., Rettenmeier A.W., Hirner A.V., Obe G. (2004). Uptake of inorganic and organic derivatives of arsenic associated with induced cytotoxic and genotoxic effects in Chinese hamster ovary (CHO) cells. *Toxicology and Applied Pharmacology*, **201**(2):156-65.

- European Food Safety Authority (EFSA). Panel on Contaminants in the Food Chain (CONTAM). Scientific Opinion on Arsenic in Food. EFSA Journal, 2009; **7**(10):1351.
- Ferreccio C., Gonzalez C., Milosavjlevic V., Marshall G., Sancha A.M., Smith A.H. (2000). Lung cancer and arsenic concentrations in drinking water in Chile. *Epidemiology*; **11**:673–79.
- Goessler W., Kuehnelt D. (2002). Analytical methods for the determination of arsenic and arsenic compounds in the environment. *Environmental Chemistry of Arsenic*, 27-50.
- Hirano S., Cui X., Li S., Kanno S., Kobayashi Y., Hayakawa T., Shraim A. (2003). Difference in uptake and toxicity of trivalent and pentavalent inorganic arsenic in rat heart microvessel endothelial cells. *Archives of Toxicology*, **77**(6):305-12.
- Hopenhayn-Rich C., Biggs M.L., Smith A.H. (1998). Lung and kidney cancer mortality associated with arsenic in drinking water in Cordoba, Argentina. *Int J Epidemiol*; **27**:561–69.
- Kitchin K.T. (2001). Recent advances in arsenic carcinogenesis: Modes of action, animal model systems, and methylated arsenic metabolites. *Toxicology and Applied Pharmacology*, **172**(3):249-61.
- Kolozsi A., Lakatos A., Galbács G., Madsen A.Ø., Larsen E., Gyurcsik B. (2008). A pH-metric, UV, NMR, and X-ray crystallographic study on arsenous acid reacting with dithioerythritol. *Inorganic Chemistry*, **47**(9):3832-40.
- Kozul C.D., Ely K.H., Enelow R.I., Hamilton J.W. (2009). Low-dose arsenic compromises the immune response to influenza a infection in vivo. *Environ Health Perspect*; **117**(9):1441-47.
- Lin S., Del Razo L.M., Styblo M., Wang C., Cullen W.R., Thomas D.J. (2001). Arsenicals inhibit thioredoxin reductase in cultured rat hepatocytes. *Chemical Research in Toxicology*, **14**(3):305-311.
- Mass M.J., Tennant A., Roop B.C., Cullen W.R., Styblo M., Thomas D.J., Kligerman A.D. (2001). Methylated trivalent arsenic species are genotoxic. *Chemical Research in Toxicology*, **14**(4):355-61.
- Meharg A.A., Hartley-Whitaker J. (2002). Arsenic uptake and metabolism in arsenic resistant and nonresistant plant species. *New Phytologist*, **154**(1):29-43.
- Murcott S. (2012). Arsenic Contamination in the World An International Sourcebook *IWA monograph*. ISBN: 9781780400389, Pages: 344.
- Naranmandura H., Suzuki N., Suzuki K.T. (2006). Trivalent arsenicals are bound to proteins during reductive methylation. *Chemical Research in Toxicology*, **19**(8):1010-18.
- Naranmandura H., Ibata K., Suzuki K.T. (2007). Toxicity of dimethylmonothioarsinic acid toward human epidermoid carcinoma A431 cells. *Chemical Research in Toxicology*, 20(8):1120-25.

- Ng J.C. (2005). Environmental contamination of arsenic and its toxicological impact on humans. *Environmental Chemistry*, **2**(3):146-60.
- Petrick J.S., Ayala-Fierro F., Cullen W.R., Carter D.E., Vasken Aposhian H. (2000). Monomethylarsonous acid (MMA(III)) is more toxic than arsenite in Chang human hepatocytes. *Toxicology and Applied Pharmacology*, **163**(2):203-7.
- Raab A., Wright S.H., Jaspars M., Meharg A.A., Feldmann J. (2007). Pentavalent arsenic can bind to biomolecules. *Angewandte Chemie - International Edition*, **46**(15):2594-97.
- Raml R., Rumpler A., Goessler W., Vahter M., Li L., Ochi T., Francesconi K.A. (2007). Thio-dimethylarsinate is a common metabolite in urine samples from arsenic-exposed women in Bangladesh. *Toxicology and Applied Pharmacology*, **222**(3):374-80.
- Rey N.A., Howarth O.W., Pereira-Maia E.C. (2004). Equilibrium characterization of the As(III)-cysteine and the As(III)-glutathione systems in aqueous solution. *Journal of Inorganic Biochemistry*, **98**(6):1151-59.
- Schuliga M., Chouchane S., Snow E.T. (2002). Upregulation of glutathione-related genes and enzyme activities in cultured human cells by sublethal concentrations of inorganic arsenic. *Toxicological Sciences*, **70**(2):183-92.
- Shiobara Y., Ogra Y., Suzuki K.T. (2001). Animal species difference in the uptake of dimethylarsinous acid (DMAIII) by red blood cells. *Chemical Research in Toxicology*, **14**(10):1446-52.
- Smith A.H., Hopenhayn-Rich C., Bates M.N., Goeden H.M., Hertz-Picciotto I., Duggan H.M., Wood R., Kosnett M.J., Smith M.T. (1992). Cancer risks from arsenic in drinking water. *Environmental Health Perspectives*, **97**:259-67.
- Smith A.H., Goycolea M., Haque R., Biggs M.L. (1998). Marked increase in bladder and lung cancer mortality in a region of northern Chile due to arsenic in drinking water. *Am J Epidemiol*;**147**(7):660–69.
- Spuches A.M., Kruszyna H.G., Rich A.M., Wilcox D.E. (2005). Thermodynamics of the as(III)-thiol interaction: Arsenite and monomethylarsenite complexes with glutathione, dihydrolipoic acid, and other thiol ligands. *Inorganic Chemistry*, **44**(8):2964-72.
- Styblo M., Del Razo L.M., Vega L., Germolec D.R., LeCluyse E.L., Hamilton G.A., Reed W., Wang C., Cullen W.R., Thomas D.J. (2000). Comparative toxicity of trivalent and pentavalent inorganic and methylated arsenicals in rat and human cells. *Archives of Toxicology*, **74**(6):289-99.
- Suzuki K.T., Tomita T., Ogra Y., Ohmichi M. (2001). Glutathione-conjugated arsenics in the potential hepato-enteric circulation in rats. *Chemical Research in Toxicology*, **14**(12):1604-11.
- Suzuki K.T., Iwata K., Naranmandura H., Suzuki N. (2007). Metabolic differences between two dimethylthioarsenicals in rats. *Toxicology and Applied Pharmacology*, **218**(2):166-73.
- Suzuki N., Naranmandura H., Hirano S., Suzuki K.T. (2008). Theoretical calculations and reaction analysis on the interaction *of pentavalent thioarsenicals with* biorelevant thiol compounds. *Chemical Research in Toxicology*, **21**(2):550-3.

- Vallee B.L., Ulmer D.D., Wacker W.E.C. (1960). Arsenic *toxicology and biochemistry. Arch. Ind. Health*, **21**:132-51.
- Wang T.S., Shu Y.F., Liu Y.C., Jan K.Y., Huang H. (1997). Glutathione peroxidase and catalase modulate the genotoxicity of arsenite. *Toxicology*, **121**(3):229-37.
- WHO (World Health Organization). (2011). Guidelines for Drinking-water Quality. 4th ed. WHO, Geneva, Switzerland. ISBN 978 92 4 154815 1.
- Wu M.M., Kuo T.L., Hwang Y.H., Chen C.J. (1989). Dose-response relation between arsenic concentration in well water and mortality from cancers and vascular diseases. *Am J Epidemiol*; **130**(6):1123–32.
- Zahler W.L., Cleland W.W. (1968). A specific and sensitive assay for disulfides. *Journal of Biological Chemistry*, **243**(4):716-19.

Chapter 10
Barium (Ba)

Adriana Floridia, Fiore Maria

10.1 ENVIRONMENTAL EFFECT

Barium, atomic number 56, is a dense alkaline earth metal in Group IIA of the periodic table. The free element is a silver-white soft metal that takes on a silver-yellow color when exposed to air (ASTDR, 2007). Barium does not exist in nature in the elemental form but occurs as the divalent cation in combination with other elements (EPA, 1998). It is present as a trace element in both igneous and sedimentary rocks. Although it is not found free in nature, it occurs in a number of compounds, most commonly barium sulfate (barite or Barytes) and, to a lesser extent, barium carbonate (WHO, 2011). These compounds are solids, existing as powders or crystals, and they do not burn well. Barium enters the environment through the weathering of rocks and minerals and through anthropogenic releases (EPA, 1998).

Barium and barium compounds are used for many important purposes. Barium sulfate ore is mined and used in several industries. It is used mostly by the oil and gas industries to make drilling muds. Drilling muds make it easier to drill through rock by keeping the drill bit lubricated. Barium sulfate is also used to make paints, bricks, tiles, glass, rubber, and other barium compounds. Some barium compounds, such as barium carbonate, barium chloride, and barium hydroxide, are used to make ceramics, insect and rat poisons, and additives for oils and fuels; in the treatment of boiler water; in the production of barium greases; as a component in sealants, paper manufacturing, and sugar refining; in animal and vegetable oil refining; and in the protection of objects made of limestone from deterioration. Barium sulfate is sometimes used, as a radiopaque contrast compound, by doctors to perform medical tests and take x-ray photographs of the stomach and intestines (ASTDR, 2007).

Background levels of barium in the environment are very low. The air that most people breathe contains about 0.0015 parts of Ba per billion parts of air (ppb). The air around factories that release barium compounds into the air has about 0.33 ppb or less of barium. Most surface water and public water supplies contain on average 0.030 parts of barium per million parts of water (ppm) or less, but can average as high as 0.30 ppm in some regions of the United States. People with the greatest known risk of exposure to high levels of barium are those working in industries that make or use barium compounds (ASTDR, 2007). The primary route of exposure to barium appears to be ingestion from food and drinking water. Barium in water comes primarily from natural sources. The highest levels to be found in drinking-water are likely to be associated with groundwater of low pH from granite-like igneous rocks, alkaline igneous and volcanic rocks and manganese-rich sedimentary rocks. Concentrations are, therefore, expected to be relatively stable. The mean daily intake of barium from food, water and air is estimated to be slightly more than 1 mg/day, food being the primary source for the non-occupationally exposed population. However, where barium levels in water are high, drinking-water may contribute significantly to barium intake (WHO,

2011). Some foods, such as Brazil nuts, seaweed, fish, and certain plants, may contain high amounts of barium. The amount of barium found in food and water usually is not high enough to be a health concern. However, information is still being collected to determine if long-term exposure to low levels of barium causes any health problems (ASTDR, 2007).
Barium is a naturally occurring constituent of drinking-water and can be controlled only by source selection or drinking-water treatment. Precipitation softening and ion exchange softening are the only treatment processes capable removing a substantial proportion (>90%) of barium from drinking-water (WHO, 2011).
The guideline value of WHO for Barium in drinking-water is 0.7 mg/l (WHO, 2011).

10.2 EFFECT ON HUMAN HEALTH

Barium is not considered to be an essential element for human nutrition. At high concentrations, barium causes vasoconstriction by its direct stimulation of arterial muscle, peristalsis as a result of the violent stimulation of smooth muscles and convulsions and paralysis following stimulation of the central nervous system. Depending on the dose and solubility of the barium salt, death may occur in a few hours or a few days. The acute toxic oral dose is between 3 and 4g (WHO, 2011).
Investigations of chronic barium toxicity in humans have focused on cardiovascular toxicity, with a specific emphasis on hypertension which has not been identified.
Associations between the barium content of drinking-water and mortality from cardiovascular disease have been observed in several ecological epidemiological studies. There were no significant evidence of any health effects associated with the normally low levels of barium in water.
There is no evidence that barium is carcinogenic or mutagenic. barium has been shown to cause nephropathy in laboratory animals, but the toxicological end-point of greatest concern to humans appears to be its potential to cause hypertension (WHO, 2011).

References:

- EPA. (1998). Toxicological review of barium and compounds. U.S. Environmental Protection Agency.
- WHO (World Health Organization). (2011). Guidelines for Drinking-water Quality. 4th ed. WHO, Geneva, Switzerland. ISBN 978 92 4 154815 1.
- ASTDR. (2007). Toxicological profile for Barium and Barium Compounds. U.S. Department of Health and Human Services.

Chapter 11
Beryllium (Be)

Maria Grazia D'Agati, Giovanni Arena.

11.1 ENVIRONMENTAL EFFECT

Beryllium, atomic number 4 (Chemical Abstracts Service Registry No. 7440-41-7; glucinium), is a inorganic metallic element. It is steel-grey, brittle metal atomic weight of 9.01 (Group IIA of the periodic table) (WHO, 2011).

Beryllium is an element that occurs naturally. It is present in a variety of materials, such as rocks, coal and oil, soil, and volcanic dust. Two kinds of mineral rocks, bertrandite and beryl, are mined commercially for the recovery of beryllium. Very pure gem-quality beryl is better known as either aquamarine (blue or blue-green) or emerald (green). Beryllium is the lightest metal. A key distinction among beryllium compounds is that some are soluble in water, but many are not (ATSDR, 2008).

Most of the beryllium ore that is mined is converted into alloys (mixtures of metals). Most of these alloys are used in making electrical and electronic parts or as construction materials for machinery and molds for plastics. Beryllium alloys are also used in automobiles, computers, sports equipment (such as golf clubs and bicycle frames), and dental bridges. Pure beryllium metal is used in nuclear weapons and reactors, aircraft and space vehicle structures, instruments, x-ray machines, and mirrors. Beryllium oxide is also made from beryllium ores and is used to make specialty ceramics for electrical and high-technology applications (ATSDR, 2002; Ross et al., 2009)

The primary source of beryllium compounds in water appears to be release from combustion of coal and industrial use of beryllium. Other sources of beryllium in surface water include deposition of atmospheric beryllium and weathering of rocks and soils containing beryllium. In most natural waters, the majority of beryllium will be adsorbed to suspended matter or in sediment, rather than dissolved. Beryllium is not likely to be found in natural water above trace levels as a result of the insolubility of oxides and hydroxides at the normal pH range. Surface waters have been reported to contain beryllium at concentrations up to 1000 ng/l (WHO, 2009).

The general population may be exposed to trace amounts of beryllium by inhalation of air, consumption of drinking-water and food, and inadvertent ingestion of dust. The estimated total daily beryllium intake in the USA was 423 ng, with the largest contributions from food (120 ng/day, based on daily consumption of 1200 g of food containing a beryllium concentration of 0.1 ng/g fresh weight) and drinking-water (300 ng/day, based on daily intake of 1500 g of water containing beryllium at 0.2 ng/g), with smaller contributions from air (1.6 ng/day, based on daily inhalation of 20 m^3 of air containing a beryllium concentration of 0.08 ng/m^3) and dust (1.2 ng/day, based on daily intake of 0.02 g/day of dust containing beryllium at 60 ng/g) (WHO, 2009).

As beryllium is rarely, if ever, found in drinking-water at concentrations of concern, it is not considered necessary to set a formal guideline value (WHO, 2011).

11.2 EFFECT ON HUMAN HEALTH

The general population can be exposed to beryllium via inhalation, oral, and dermal routes of exposure. The inhalation route is of greatest concern for systemic effects because beryllium and its compounds are poorly absorbed after oral and dermal exposure. The respiratory tract in humans and animals is the primary target of beryllium toxicity following inhalation exposure (USACHPPM, 1998). Occupational exposure to high concentrations of soluble beryllium compounds can result in acute beryllium disease, while exposure to relatively low concentrations of soluble or insoluble beryllium compounds can result in chronic beryllium disease. Acute beryllium disease is characterized by inflammation of the respiratory tract tissues, pneumonitis, cough, chest pain, dyspnea, and pneumonia and is usually resolved within several months of exposure termination (Aw et al., 2007). In contrast, chronic beryllium disease is an immune response to beryllium and is only observed in individuals who are sensitized to beryllium (usually <15% of an exposed population) (Richard et al., 2011). Chronically, it results in sarcoid-like granulomata mainly in the lungs and is occasionally subcutaneous. These granulomas result from an immune reaction, primarily based on cell-mediated immunity (WHO, 2009). A genetic component to chronic beryllium disease susceptibility has also been identified (Marchand et al., 2005).

Beryllium is a likely candidate as a lung carcinogen (Aw et al., 2007; Kuschner, 1981). Beryllium compounds cause genetic transformations in cultured mammalian cells resulting from binding of ionic beryllium to nucleic acids, resulting in infidelity of DNA replication (Gordon et al., 2003; Strupp, 2011). Other systemic effects that have been observed in individuals with severe cases of chronic beryllium disease include damage to the right heart ventricle, hepatic necrosis, kidney stones, and weight loss; these effects are probably secondary to chronic beryllium disease rather than a direct effect on the tissues.

The toxicity of beryllium compounds increases with increasing water solubility.

In the European community, beryllium and its compounds are classified as Class 2 carcinogens ('known animal carcinogens that are suspected to be human carcinogens') (Strupp et al., 2011).

A number of epidemiology studies have been conducted to assess the carcinogenic potential of beryllium. Increased incidence of lung cancer deaths were reported in retrospective cohort mortality studies of workers at beryllium extraction, processing, and fabrication facilities. Increased lung cancer mortality was also seen in entrants to the Beryllium Case Registry. No correlation between the incidence of lung cancer deaths and exposure has been established because historical exposure levels were not reported. A positive association between length of latency and lung cancer deaths was found, with the highest cancer risks among workers with a latency of 25 years (ATSDR 2002; Gordon et al., 2003).

References:

- Agency for Toxic Substances and Disease Registry (ATSDR). (2008). Public health statement. Beryllium. Cas#: 7440-41-7. Buford Highway, Atlanta, Georgia, USA
- Agency for Toxic Substances and Disease Registry (ATSDR). (2002). Toxicological Profile for Beryllium. Atlanta, Georgia, USA

- Aw T.C., Gardiner K., Harrington J.M. (2007). Pocket Consultant. Occupational Health. 5th ed. Oxford: Blackwell Publishing; Chapter 5. Occupational Toxicology.
- Cooper R.G., Harrison A.P. (2009). The uses and adverse effects of beryllium on health. *Indian J Occup Environ Med*; **13**(2): 65–76.
- Gordon T., Bowser D. (2003). Beryllium: genotoxicity and carcinogenicity. *Mutation Research,* **533**:99–105.
- Kuschner M. (1981). The carcinogenicity of beryllium. *Environ Health Perspect*; **40**:101–5.
- Marchand-Adam S., Guillon F., Brauner M., Valeyre D. (2005). Chronic beryllium disease: A model of interactions between environmental exposure and genetic predisposition. Pathogenesis and clinical features (Part 2). *Rev Mal Respir.*; **22**:271–87.
- Richard T., Maier S.L.A. (2011). Chronic beryllium disease: an updated model interaction between innate and acquired immunity. *Biometals,* **24**:1–17.
- Strupp C. (2011) Beryllium Metal II. A Review of the Available Toxicity Data. *Ann. Occup. Hyg.*, **55**(1):43–56.
- USACHPPM. (1998). Beryllium Exposure and Berylliosis. Maryland, USA: Occupational and Environmental Medicine, US Army Centre for Health Promotion and Preventative Medicine.
- WHO (World Health Organization). (2009). Background document for development of WHO Guidelines for Drinking-water Quality: Beryllium in drinking-water. Geneva.
- WHO (World Health Organization). (2011). Guidelines for Drinking-water Quality. 4th ed. WHO, Geneva, Switzerland. ISBN 978 92 4 154815 1.

Chapter 12
Bismuth (Bi)

Cristina Mauceri, Chiara Copat.

12.1 ENVIRONMENTAL EFFECT

Bismuth, atomic number 83, is a white, crystalline, brittle metal with a pinkish tinge. It has a high electrical resistance and features the highest diamagnetic properties. The most important ores of bismuth are bismuthimite and bismite. All bismuth salts form insoluble compounds when put into water.
For it, not was considered necessary to set a formal guideline value .
The abundance of bismuth in the Earth's crust is estimated to be about 0.2 parts per million, making it a relatively rare element Bismuth occurs naturally as metal itself and it is found as crystals in the sulphides ores of nickel, cobalt, silver and tin. Bismuth is mainly produced as a by-product from lead and copper smelting. Bismuth metal is used in the manufacture of low melting solders and fusible alloys as well as low toxicity bird shot and fishing sinkers. Certain bismuth compounds are also manufactured and used as pharmaceuticals for the treatment of gastrointestinal problems like gastritis, peptic ulcer and diarrhoea (WHO). Industry makes use of bismuth compounds as catalysts in manifacturing acrylonitrile, the starting material for synthetic fibers and rubbers. Bismuth is sometimes used in the production of shot and shotguns. There is increasing interest in using bismuth as a substitute for lead in alloys since lead is toxic to humans and animals. Bismuth compounds are used in medications. Bismuth can enter the body through three means - inhalation, ingestion and through skin absorption. The most common exposures to bismuth include working in close correlation with the element.

12.2 EFFECT ON HUMAN HEALTH

Bismuth and its compounds are not considered toxic elements. However, when exposure to bismuth is elevated, there are a number of health issues that may arise, and if the exposure level is severe enough, the results can be even fatal.
When high levels of bismuth come in contact with the skin or eyes, it can cause some irritation at the site. Ingestion and inhalation are the two exposure methods that can carry the most dangerous effects, including foul breath, metallic taste and gingivitis nausea, loss of appetite and weight, malaise, albuminuria, diarrhoea, skin reactions, stomatitis, headache, fever, sleeplessness, depression, rheumatic pain a black line may form on gums due to deposition of bismuth sulphide. Chronic exposure may affect the function of the liver and the kidneys and may cause anemia. Bismuth is not considered a human carcinogen.

References:

- www.globalhealingcenter.com (accessed on 9 September 2012).
- www.lenntech.com (accessed on 9 September 2012).
- www.chemistryexplained.com (accessed on 10 September 2012).
- WHO. (World Health Organization). Basic Analytical Toxicology of Bismuth. Monographs - analytical and toxicological data (6.12 - 6.23). http://www.who.int/ipcs/publications/training_poisons/basic_analytical_tox/en/index8.html (accessed on 10 September 2012).

Chapter 13
Boron (B)

Gea Oliveri Conti.

13.1 ENVIRONMENTAL EFFECT

Boron is a chemical element with symbol B and atomic number 5. Boron is a rare element in the Earth, representing only 0.001%.
Chemically uncombined boron, which is classed as a metalloid, is not found naturally on Earth. Industrially, very pure boron is produced with difficulty, as boron tends to form refractory materials containing small amounts of carbon or other elements.
In biology the borates have low toxicity in mammals but are most toxic to arthropods, so, these are used as insecticides. Boric acid is mildly antimicrobial and the Boron is essential to life. Small amounts of boron compounds play a strengthening role in the cell walls of all plants, making boron necessary in soils. Experiments indicate a role for boron as an ultratrace element in animals, but its role in animal physiology is still unknown (LANL, 2008). Boron compounds are used in the manufacture of glass, soaps, detergents, electronics and as flame retardants. Naturally occurring boron is present in groundwater primarily as a result of leaching from rocks and soils containing borates and borosilicates. High levels of boron concentrations were found mostly in rivers that are associated with urban and industrial drainage relative to those from rural areas, thus indicating a direct link between boron contamination in rivers and sewage pollution, but use of boron has decreased significantly, so, levels of boron in wastewater discharges continue to fall.

13.2 EFFECT ON HUMAN HEALTH

Boron is absorbed almost completely by gastrointestinal and respiratory systems and mostly present in body tissues and fluids as Boric acid. Boric acid is known as a natural compound and it is taken from water, vegetables and many other foods through diet (WHO, 2011). In many populations food ingredients and occasionally drinking water of volcanic areas are the major source of boron intake (Bakirdere et al., 2010).
The amount of boron intake by people is dependent on the types of food consumed e.g. fruits, vegetables with leaves and legumes contain relatively high amounts of boron. Boron concentration in meat, fish and dairy food products is not high. Coffee and milk contain low amount of boron. Hazelnut butter, avocado, vine, peanuts and other nuts with shells have high amounts of boron. According to WHO, it is expected that the exposure to boron via drinking water is 0.2-0.6 mg/day via and 1.2 mg/day via diet.
Risk assessments of boron indicate no significant risk of toxicity to humans at currently estimated dietary or municipal drinking water levels of exposure but boric acid and sodium borates have low acute toxicity (WHO, 2011). Boron compounds are toxic to all species tested at high doses, but they are not mutagenic or carcinogenic, indeed, the major chronic toxicities are developmental and reproductive (Bakirdere et al., 2010; WHO, 2011).

The boron exposure in pregnancy include many developmental effect as a high prenatal mortality, reduced fetal body weight, abnormality of cardiovascular system and central nervous system, malformations of the eyes and axial skeleton. In animals, the main toxic effect associated with boron involves the reproductive system. Boron caused specific adverse effects in the male reproductive tract in rats, mice and dogs, including shrunken scrota, inhibited spermiation, degeneration, and atrophy of seminiferous tubules, with an absence or loss of germ cells. Boron also caused a reduction in ovulation in female rats and renal lesions in female mice (WHO, 2011).

There is no data in particular regarding the ability of boron to cross the placenta or accumulate in fetal tissues. However data from animal studies suggests that developmental toxicity may be an area of concern in human's health.

Still no evidence was found to suggest that boron interferes with human fertility and reproduction. In fact a study completed with a highly exposed population in Turkey, where exposure occurs mainly from naturally high levels of boron in drinking water (up to 29 mg/l) as well as from mining and production, no adverse effects were reported on fertility over three generations, similarly, some birth-ratio study showed no evidence.

A non statistically significant increase was detected in the percentage of female offspring but this increase of the percentage of female offspring does not appear to be related to exposures to inorganic borates (Bakirdere et al., 2010).

The data on human studies are still insufficient to determine whether boron exposure may cause toxicity on male or female reproduction.

Boron is a trace element with a very important role in mineral and hormonal metabolisms, in fact it is an important cofactor in cell membrane functions and enzyme reactions. Boron also affects osteoporosis,heart trouble, paralysis, diabetes and senility. Boron is involved in calcium and bone metabolism. Some studies show that its effects are more marked when cholecalciferol and magnesium are deficient. Boron may be involved in cerebral function due to its effects on the transport across membranes. Boron is involved in the synthesis of extracellular matrix and is helpful in wound healing. Compounds of boron have been shown to be potent anti-osteoporotic, anti-inflammatory, hypolipemic, anti-coagulant, and anti-neoplastic agents both in vitro and in vivo in animals (Bakirdere et al., 2010).

For all these reasons boron is still a very controversial element.

Most recently, in 2000 the EU passed the Water Framework Directive an overarching piece of legislation that brings together all existing EU legislation on water resources (Article 4). In 1998 the EU had revised its Drinking Water Directive (98/93/EC), which is responsible for regulating the quality of water that is intended for human consumption. This Drinking Water Directive requires that the EU member states comply with a number of water quality standards and health parameters, and in the 1998 Drinking Water Directive revisions the EU added several new chemical parameters one of which is a parametric value for the element boron in drinking water (1 mg/l) . The boron directive largely affects the European member states in the Mediterranean basin (e.g. Cyprus, Greece, Italy, and Spain) because some of their groundwater bodies contain exceptionally high levels of boron (Weinthal et al., 2005).

EU legislation has focused on the precautionary approach as a mechanism to contend with scientific uncertainty concerning the health effects of a potential pollutant . Instead will be most important shift attention to the way in which scientific uncertainty about the causes of a pollutant can impede compliance with new regulatory standards.

The new boron standard, in particular, can create subsequent compliance problems for EU member states in developing timely and economically viable management solutions (Weinthal et al., 2005).
The guideline value of WHO for Boron in drinking-water is 2.4 mg/l (WHO, 2011).

References:

- Bakirdere S., Örenay S., Korkmaz M. (2010). Effect of Boron on Human Health. *The Open Mineral Processing Journal*, **3**:54-9.
- LANL. Los Alamos National Laboratory. (2008) . Boron Properties. *Retrieved*, 9-18.
- Weinthal E., Parag Y., Vengosh A. Muti A., Kloppmann W.. (2005). *The EU drinking water directive: the boron standard and scientific uncertainty*. Eur. Env. **15**:1–12.
- WHO (World Health Organization). (2011). Guidelines for Drinking-water Quality. 4th ed. WHO, Geneva, Switzerland. ISBN 978 92 4 154815 1.

Chapter 14
Calcium (Ca)

Maria Grazia D'Agati, Caterina Ledda.

14.1 ENVIRONMENTAL EFFECT

The element Calcium, atomic number 20, is the third most abundant metal in the earth crust. Calcium compounds account for 3.64% of the earth crust. The distribution of calcium is very wide as it is found in almost every terrestrial area. This element is essential for the life of plants and animals, for it is present in endo- and exo- skeletons, in teeth, in egg shells, in coral and in many soils. Seawater contains 0.15% of calcium chloride.
Calcium cannot be found alone in nature; it is found mostly as limestone, gypsum and fluorite. Stalagmites and stalactites contain calcium carbonate. Calcium ions dissolve in water form deposits in pipes and boilers and when the water is hard, that is, when it contains too much calcium or magnesium. This can be avoided through water softeners. In contact with air, calcium develops an oxide and nitride coating, that protects it from further corrosion. It burns in the air at high temperature to produce nitride.
Calcium occurs in water naturally. Seawater contains approximately 400 ppm of calcium. One of the main reasons for the abundance of calcium in water is its natural occurrence in the earth crust. Inan aqueous solution calcium is mainly present as Ca^{2+} (aq), but it may also occur as $CaOH^{+}$ (aq) or $Ca(OH)_2$ (aq), or as $CaSO_4$ in seawater. Calcium functions as a pH stabilizer, because of its buffering qualities, and it also gives water a better taste. Calcium is a dietary requirement for all organisms apart from some insects and bacteria. Calcium carbonate is a building stone of skeletons of most marine organisms, and eye lenses. Calcium phosphate is required for bone structure and teeth structure of terrestrial organisms. Plants mainly contain calcium oxalate. Calcium storage in plants is about 1% of dry mass. Calcium is largely responsible for water hardness, and may negatively influence toxicity of other compounds. Elements such as copper, lead and zinc are much more toxic in soft water. Water hardness influences aquatic organisms concerning metal toxicity. In softer water membrane permeability in the gills is increased. Calcium also competes with other ions for binding spots in the gills. Consequently, hard water better protects fishes from direct metal uptake. pH values of 4.5-4.9 may harm salmon eggs and grown salmons, when the calcium, sodium and chlorine content is low. Various calcium compounds may be toxic. Calcium carbide forms flammable acetylene when it comes in contact with water and is therefore considered hazardous. Some environmental effects of water hardness include hardening of domestic equipment, because high temperatures cause carbonate hardness. This may dramatically decrease the lifespan of equipment, and causes an increase of domestic waste.
Calcium carbonate interacts with detergents and cleansing agents. Complex formation causes a decrease in detergent efficiency, resulting in increased detergent application and softener purchases. Softening is often carried out by means of ion exchangers. These must be regenerated with kitchen salt, and therefore they burden wastewater. There are six

stable isotopes of calcium. Today, we know of eight unstable Calcium isotopes. ^{45}Ca is highly radioactive and toxic.

Drinking water is not the major source of our calcium intake; the health significance of supplemental intake of this element from drinking water may outweigh its nutritional contribution expressed as the proportion of the total daily intake of this element. However, drinking-water may be an important source of mineral intake; that is why the shift in consumption from tap water to bottled water may have important implications in health and disease (WHO; 2005).

Drinking water supplies may contain some essential minerals naturally or through deliberate or incidental addition. Water supplies are highly variable in their mineral contents and, while some contribute appreciable amounts of certain minerals either due to natural conditions (e.g., Ca, Mg, Se, F, Zn), intentional additions (F), or leaching from piping (Cu), most provide lesser amounts of nutritionally-essential minerals. Many persons consume mineral waters because of the perception that they may be healthier.

Vitamin D is an important co-factor in the intestinal absorption of calcium, as it increases the number of calcium binding proteins, involved in calcium absorption through the apical membrane of enterocytes in small intestine. It also promotes re-absorption of calcium in the kidneys

The contribution of water to calcium intake depends on the amount of the minerals in the water and the amount of water consumed. The variation in the mineral content of bottled waters in different parts of the world is tremendous, and few of them have an optimal mineral profile (WHOa; 2005).

The potential contributions of drinking water to nutritional status also depend on behavioral factors and environmental conditions. Individuals with the greatest relative consumption of water include infants, residents in hot climates, and individuals engaged in strenuous physical activity. Health consequences of micronutrient deficiencies include increased morbidity, mortality due to reduced immune defense systems and impaired physical and mental development (WHOb; 2005).

14.2 EFFECT ON HUMAN HEALTH

Since the early 1960's, epidemiological studies in many countries all over the world have reported that soft water (i.e., water low in calcium) is associated with increased morbidity and mortality from cardiovascular disease (CVD) compared to hard water. Recent studies also suggest that the intake of soft water may be associated with higher risk of fracture in children (Verd Vallespir et al., 1992), certain neurodegenerative diseases, pre-term birth and low weight at birth (Yang et al., 2002) and some types of cancer (Yang et al., 1997; Yang et al., 1998). The importance of calcium in water was also confirmed in a one-year study of rats on a fully adequate diet in terms of nutrients and salts; rats were given desalinated water with added dissolved solids of 400 mg/l and either 5 mg/l, 25 mg/l, or 50 mg/l of calcium. The animals given water dosed with 5 mg/l of calcium exhibited a reduction in thyroidal and other associated functions compared to the animals given the two higher doses of calcium (Kozisek, 2004).

Calcium is the most abundant metal in the human body (see Fig. 14.1); it is the main constituent of bones and teeth and it has keys metabolic functions. Calcium plays an important role in the human body as an essential bone mineral as well as a messenger in signal transduction (Bushinsky & Monk RD, 1998). Its concentration is regulated by a variety of hormones and colecalciferol. In principle, the healthy body is able to eliminate excessive calcium, but is unable to cope with low calcium intake for a extended period of time (Bohmer et al., 2000).

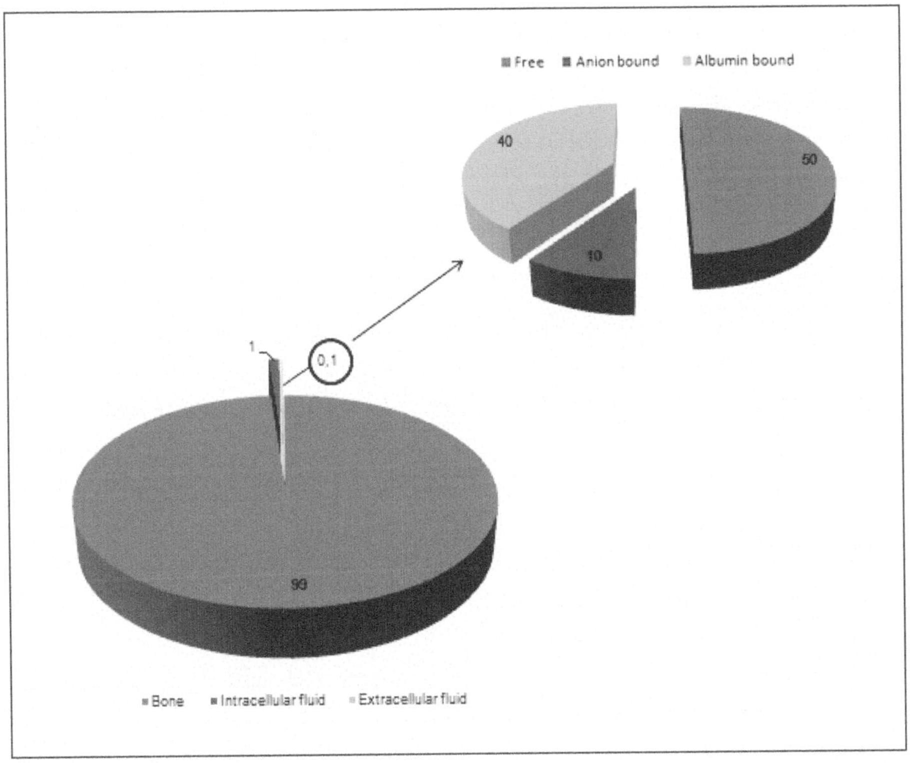

Fig. 14.1: Ca content (%) in human body.

Inadequate intakes of calcium have been associated with increased risks of osteoporosis, nephrolithiasis (kidney stones), colorectal cancer, hypertension and stroke, coronary artery disease, insulin resistance and obesity. Most of these disorders have treatments but no cures. Owing to a lack of compelling evidence for the role of calcium as a single contributory element in relation to these diseases, estimates of calcium requirement have been made on the basis of bone health outcomes, with the goal of optimizing bone mineral density. Calcium is unique among nutrients, in that the body's reserve is also functional: increasing bone mass is linearly related to reduction in fracture risk.

A sufficient calcium intake is now considered to be beneficial to bone mass at all stages of life and some recommended dietary allowances (RDAs) for calcium are between 1200 and 1500 mg/d for adolescents, pregnant and nursing women, and people aged 65 (NIH, 1994). To attain the optimal calcium intake it has been suggested that the frequency of consumption of dairy products and calcium-rich vegetables be increased. Other low-energy calcium sources, not mentioned in the National Institutes of Health Consensus Statement, are high-calcium mineral waters. The bioavailability of calcium from mineral waters was recently studied and was found to be at least equal to that from milk (Heaney, 2006).

Other studies show that high-calcium mineral water not only represented an additional dietary source of calcium but also modulated parathyroid function and bone metabolism. Nevertheless, further long-term studies are needed to confirm these short-term results (Guillemant et al., 2000).

Between 1 and 2l of a typical calcium-rich mineral water with a concentration of about 500 mg calcium/l can help satisfy even an increased demand. This recommendation is based on the assumption of an adequate bioavailability of the ingested calcium. Although only a few studies with a relatively small number of subjects are available, the bioavailability of calcium from Ca-rich mineral waters seems to be at least comparable to, and possibly better than, that from dairy products or pharmaceutical preparations. Mineral waters are not mono-substances. Often different minerals are leached in relevant concentrations. Available study results indicate that calcium absorption is not substantially influenced by other accompanying ion species (Bohmer et al., 2000).

References:

- Bohmer H., Muller H., Resch K.L. (2000). Calcium supplementation with calcium-rich mineral waters: a systematic review and meta- analysis of its bioavailability. *Osteoporosis Int*, **11**:938-43. International Osteoporosis Foundation and National Osteoporosis Foundation.
- Bushinsky D.A., Monk R.D. (1998). Electrolyte quintet: calcium. Lancet, **352**:306-11.
- Kozisek F. (2004). Rolling revision of the WHO Guidelines for drinking-water quality Draft for review and comments. Health risks from drinking demineralised water. World Health Organization Geneva.
- Guillemant J., Le H.T., Accarie C., Tézenas du Montcel S., Delabroise A.M., Arnaud M.J., Guillemant S. (2000). Mineral water as a source of dietary calcium: acute effects on parathyroid function and bone resorption in young men. *Am J Clin Nutr*; **71**:999–1002.
- Heaney R.P. (2006). Absorbability and utility of calcium in mineral waters. *Am. J. Clin. Nutr.* **84**:371–74.
- Yang Ch.Y., Chiu H.F., Chang Ch., Wu T.N., Sung F.Ch. (2002). Association of very low birth weight with calcium levels in drinking water. *Environ. Research, Section A* **89:**189-94.
- Yang Ch.Y., Cheng M.F., Tsai S.S., Hsieh Y.L. (1998). Calcium, magnesium, and nitrate in drinking water and gastric cancer mortality. *Jpn. J. Cancer Res.* **89**:124-30.
- Yang Ch.Y., Chiu H.F., Chiu J.F., Tsai S.S., Cheng M.F. (1997). Calcium and magnesium in drinking water and risk of death from colon cancer. *Jpn. J. Cancer Res.* **88**:928-33.
- NIH (1994). National Institutes of Health. Consensus Conference. Optimal calcium intake. *JAMA*; **272**:1942–8.
- Verd Vallespir S., Domingues Sanches J., Gonzales Quintial M., Vidal Mas M., Mariano Soler A.C., de Roque Company C., Sevilla Marcos J.M. (1992). Calcium

- Association between calcium content of drinking water and fractures in children (in Spanish). *An. Esp. Pediatr.* **37**:461-65.
- WHOa. (World Health Organization). (2005). Calcium and Magnesium in Drinking-water. Geneva.
- WHOb. (World Health Organization). (2005). Nutrients in drinking water. Geneva.
- http://www.lenntech.com/periodic/elements/ca.htm.

Chapter 15
Cadmium (Cd)

Zorica Rasic-Milutinovic, Dragana Jovanovic, Ignatius Maduka.

15.1 ENVIRONMENTAL EFFECT

Cadmium is a toxic metal, atomic number 48, omnipresent in the environment. Cadmium is a soft, bluish-white metal and is easily cut with a knife. It is s imilar in many respects to zinc. Cadmium and its compounds are highly toxic.Natural and anthropogenic sources have increased cadmium levels in the biosphere as a consequence of extensive industrialization and environmental pollution. Cadmium occurs in almost all soils, surface waters and plants, and is considered a potential threat to wildlife species, including game animals. Cadmium has a very long biological halflife of 15–30 years, primarily because of its low rate of excretion from the body, and accumulates over time in the blood, kidney, and liver.

It is toxic to a wide range of organs and tissues, but the primary targets are the liver and kidney.

Occupational pollution with cadmium results mainly from metallurgy industry and manufactures of nickel–cadmium batteries, mining, pigments, coatings and plating, and as stabilizers for plastics. Silver solder, which contains cadmium, should be handled with care. Important sources of human intoxication are cigarette smoke as well as food, water and air contamination (Malaguarnera, 2012).

Cadmium contamination in drinking-water may be caused by impurities in the zinc of galvanized pipes and solders and some metal fittings, although, food is the main source of daily exposure to cadmium. Occurrence of cadmium in drinking water is usually less than 1 µg/l and the daily oral intake is 10–35 µg. The guideline value of WHO for Cadmium in drinking-water is 0.003 mg/l (WHO, 2011).

The commonest general population exposure to cadmium is from potatoes and peanuts (ground nuts) grown on Cd- contamiated soil. The origins of contamination relate to historic use of guano fertilisers which derived their Cd from shellfisn consumed by seabirds.

Cigarette smoking is a significant additional source of cadmium exposure (WHO, 2011). Biological monitoring of cadmium in the general population has shown that cigarette smoking may cause significant increases in blood Cd levels, the concentrations in smokers being on average 4–5 times higher than those in non-smokers (Jarup et al., 1998).

Cadmium is a very dangerous toxicant because of its ability to interact with many zinc finger proteins that control cellular pathways (Scandurra et al., 2012).

15.2 EFFECT ON HUMAN HEALTH

Cadmium adverse health effects after long term (see Table 15.1) include kidney dysfunction and bone damage via oxidative stress (osteomalacia, osteoporosis, fractures) (Nair et al., 2012). The first sign of the renal lesion is usually a tubular and glomerular dysfunction, evidenced by an increased excretion of low molecular weight proteins such as β_2-

microglobulin and α_1-microglobulin (human comlex-forming protein) or enzymes such as N-Acetyl-β-D-glucosaminidase (NAG) (Jarup et al. 1998; Akesson et al., 2005). The symptoms of cadmium nephrotoxicity in humans and experimental animals manifest in the defects of reabsorptive and secretory functions of proximal tubules, and include phosphaturia, aminoaciduria, glucosuria, proteinuria, increased excretion of organic anions and cations, and polyuria.

These symptoms indicate that cadmium targets various transporters in the cadmium is a toxic metal, omnipresent in the environment brush-border and basolateral membrane (Sabolić et al., 2012).

To determine critical concentration of cadmium in urine for glomerular and tubular effects the excretion of biomarker β2-microglobulin to cadmium excretion is mostly used. According to JECFA estimation, urinary excretion of less than 5.24 µg of cadmium per gram creatinine was not associated with an increased excretion of β2-microglobulin, that corresponds to dietary exposure of 0.8 µg/kg body weight per day (WHO, 2011). The lower bench-mark doses of urinary cadmium levels for the effects of kidney dysfunction were calculated using hybrid approach (Suwazono et al., 2006). Long-term high cadmium exposure may cause skeletal damage, first reported from Japan, where the itai-itai (ouch-ouch) disease (a combination of osteomalacia and osteoporosis) was discovered in the 1950s. The exposure was caused by cadmium-contaminated water used for irrigation of local rice fields (Jarup, 2003). In addition, cadmium has classified as a human carcinogen, group I, by the IARC (IARC, 1993).

Observed alterations of DNA, as consequences of cadmium applications in experimental models of mammalian cell cultures, higher plants and intact animals, include decreased fidelity of DNA synthesis, microbial DNA repair, gene mutations and chromosomal abnormalities. These positive results were often weak and seen at high concentrations that also caused cytotoxicity. However, cadmium does not appear to possess significant genotoxic potential via the oral route (WHO, 2011).

The mechanisms of carcinogenesis of cadmium are complex. They include modulation of gene expression and signal transduction, interference with enzymes of the cellular antioxidant system and generation of reactive oxygen species and DNA damage, the inhibition of different types of DNA repair and the induction of apoptosis, and role in disruption of E-cadherin-mediated cell–cell adhesion (Malaguarnera, 2012, Satarug & Moore, 2012).

Table 15.1 Cadmium effects on the body	
Organs	**Effects**
Reproductive system (Thompson & Bannigan, 2008)	Testicular necrosis, estrogen-like effects, affection of steroid-hormon synthesis.
Kidney (Nair et al., 2012)	Proteinuria, kidney stones, aminoaciduria, glucosuria, proteinuria, increased polyuria
Respiratory system (Flick et al., 1971)	Pneumonitis, destruction of mucous membranes.
Skeleton (Nair et al., 2012)	Osteomalacia, osteoporosis, fractures, Itai-Itai disease.

References:

- Akesson A., Lundh T., Vahter M., Bjellerup P., Lidfeldt J., Nerbrand C., Samsioe G., Strömberg U., Skerfving S. (2005). Tubular and glomerular kidney effects in Swedish women with low environmental cadmium exposure. *Environ Health Perspect*; **113**(11):1627–31.
- IARC. (1993). Cadmium and cadmium compounds. In: Beryllium, Cadmium, Mercury and Exposure in the Glass Manufacturing Industry. IARC Monographs on the Evaluation of Carcinogenic Risks to Humans, vol. 58. Lyon: International Agency for Research on Cancer; 119–237.
- Jarup L., Berglund M., Elinder C.G., Nordberg G., Vahter M. (1998). Health effects of cadmium exposure: a review of the literature and a risk estimate. *Scand J Work Environ Health*; **24**(Suppl 1):1–51.
- Malaguarnera M. (2012). Cadmium and cancer risk. ICS 2012, Cadmium Symposium Sassari, Italy, *Abstract book*: pp, 39.
- Nair A.R., Smeets K., Lee W.K., Thevenod F., Van Kerkhove E., Cuypers A. (2012). Cadmium induced oxidative stress in rat proximal tubular cell line: damage vs signaling. ICS 2012, Cadmium Symposium Sassari, Italy, *Abstract book*: pp, 24.
- Sabolić I., Ljubojević M., Breljak D., Herak-Kramberger C.M., Anzai N., Koepsell H. (2012). Membrane Transporters In Experimental Cadmium Nephrotoxicity. ICS 2012, Cadmium Symposium Sassari, Italy, *Abstract book*: pp, 19.
- Satarug S., Moore MR. (2012). Emerging role of cadmium and heme oxygenase in type 2 diabetes and cancer susceptibility. Tohoku J. Expt. Med. **228**:267-288
- Scandurra R., Politi L., Scotto D'Abusco A., Papa V., Pedone P., Di Luigi L., Migliaccio S. (2012). Cellular pathways altered by cadmium. ICS 2012, Cadmium Symposium Sassari, Italy, *Abstract book*: pp, 26.
- Suwazono Y., Sand S., Vahter M., Filipsson A.F., Skerfving S., Lidfeldt J., Akesson A. (2006). Benchmark Dose for Cadmium-Induced Renal Effects in Humans. *Environ Health Perspect*; **114**(7):1072–76.

- Thompson J., Bannigan J. (2008). Cadmium: toxic effects on the reproductive system and the embryo. *Reprod Toxicol*, **25**(3):304-15.
- WHO (World Health Organization). (2011). Guidelines for Drinking-water Quality. 4th ed. WHO, Geneva, Switzerland. ISBN 978 92 4 154815 1.

Chapter 16
Chromium (Cr)

Gea Oliveri Conti, Caterina Ledda, Zorica Rasic-Milutinovic, Dragana Jovanovic.

16.1 ENVIRONMENTAL EFFECT

Chromium, atomic number 24, is an odorless and tasteless metallic element, it is found naturally in rocks, plants, soil and volcanic dust, humans and animals. The most common forms of chromium that occur in natural waters in the environment are trivalent chromium or Cr (III), and hexavalent chromium or Cr(VI).

Cr (III) is an essential human dietary element and occurs naturally in many vegetables, fruits, meats, grains and yeast. Cr(VI) occurs naturally in the environment from the erosion of natural chromium deposits but it can also be produced by industrial processes. There are instances of chromium being released to the environment by leakage, poor storage, or inadequate industrial waste disposal practices (EPA, 2010; Sciacca & Oliveri Conti; 2009).

Metallic chromium is used for making steel and other alloys, and chromium compounds in either the Cr(II)I or Cr(VI) forms are used for chrome plating, dyes and pigments, leather and wood preservation (EPA, 2010).

Total chromium concentrations in drinking-water are usually less than 2 µg/l, although concentrations as high as 120 µg/l have been reported (WHO, 2011; Sciacca & Oliveri Conti; 2009).

The guideline value of WHO for total chromium in drinking-water is 0.05 mg/l (WHO, 2011).

Cr(III) is an essential dietary mineral in low doses, found in most fresh foods, including breads, meats and vegetables and drinking water (Vincent, 2010; Sciacca & Oliveri Conti; 2009).

16.2 EFFECT ON HUMAN HEALTH

Chromium is essential for normal glucose metabolism. Hexavalent Cr or Cr(VI) differs from arsenic in that it discolours water, turning the water yellow at high concentrations. Cr(VI) is well established human carcinogen via inhalation and sufficient evidence exists for its causal relationship with lung and nasal cavity cancer in humans (Quievryn et al., 2002). A controversial issue is whether chromium causes cancer when ingested (Smith & Steinmaus, 2009). Recent studyiesconducted in large populations in China found increased risk for stomach cancer mortality associated with high concentrations of Cr(VI) in drinking water (Beaumont et al., 2008; Smith & Steinmaus, 2009; Sciacca & Oliveri Conti; 2009). Other adverse effect observed in animal studies include hypotension, hepatic and renal failure, sperm abnormalities and damage to the male reproductive system (Jomova & Valko, 2011). The mutagenic potential of Cr(VI) has been studied extensively. Mutagenicity of hexavalent chromium is mediated through the generation of the highly reactive chromium intermediates Cr(V) and Cr(IV), reactive oxygen species, and Cr(III) formed during the intracellular reduction of hexavalent chromium. These free radicals that are formed during the reduction process can directly react with DNA, leading to oxidative DNA damage, Cr-DNA adducts, DNA strand breaks, and chromosomal aberrations (EPA, 2010; Sciacca & Oliveri Conti; 2009). As a marker for genotoxicity, the number of sister chromatid exchanges (SCE) in blood lymphocytes have been shown to be relatively insensitive towards exposure to Cr(VI) (Werfel et al., 1998).

References:

- Beaumont J.J., Sedman R.M., Reynolds S.D., Sherman C.D., Li L.H., Howd R.A., Sandy M.S., Zeise L., Alexeeff G.V. (2008). Cancer mortality in a Chinese population exposed to hexavalent chromium in drinking water. *Epidemiology*, **19**(1):12–23.
- EPA. (2010). IRIS. Toxicological Review of Hexavalent Chromium (External Review Draft). http://cfpub.epa.gov/ncea/iris_drafts/recordisplay.cfm?deid=221433
- Jomova K., Valko M. (2011). Advances in metal-induced oxidative stress and human disease. *Toxicol*, **283**:65–87.
- Quievryn G., Messer J., Zhitkovich A. (2002). Carcinogenic chromium(VI) induces cross-linking of vitamin C to DNA in vitro and in human lung A549 cells. *Biochemistry*; **41**:3156–67.
- Smith A.H., Steinmaus C.M. (2009). Health Effects of Arsenic and Chromium in Drinking Water: Recent Human Findings. *Annu Rev Public Health*; **29**(30):107–22.
- Vincent J.B. (2010). Chromium: celebrating 50 years as an essential element? *Dalton Trans*, **39**: 3787–94.
- Werfel U., Langen V., Eickhoff I., Schoonbrood J., Vahrenholz C., Brauksiepe A., Popp W., Norpoth K. (1998). Elevated DNA single-strand breakage frequencies in lymphocytes of welders exposed to chromium and nickel. *Carcinogenesis*, **19**(3):413-18.
- WHO (World Health Organization). (2011). Guidelines for Drinking-water Quality. 4th ed. WHO, Geneva, Switzerland. ISBN 978 92 4 154815 1.
- Sciacca S., Oliveri Conti G. (2009). Mutagens and carcinogens in drinking water. *Mediterr J Nutr Metab*, **2**:157–62.

Chapter 17
Cobalt (Co)

Adriana Floridia, Chiara Copat.

17.1 ENVIRONMENTAL EFFECT

Cobalt, atomic number 27, is a gray, hard, magnetic, ductile, and somewhat malleable metal, cobalt is relatively rare and generally obtained as a byproduct of extraction other metals such as copper (EPA).
There is only one stable isotope of cobalt, which has an atomic mass number of 59, and many unstable or radioactive isotopes, two of which are commercially important, cobalt-60 and cobalt-57. Chemically all isotopes of cobalt behave the same and will therefore have the same chemical behavior in the environment and the same chemical effects on your body (ATDSR, 2004).
Small amounts of cobalt are naturally found in most rocks, soil, water, plants, and animals. Cobalt is also found in meteorites. Cobalt does not occur naturally as a base metal, but is a component of more than 70 naturally occurring minerals, including various sulfides, arsenides, sulfoarsenides, hydrates, and oxides. The most common cobalt minerals are the arsenide CoAs2–3 (smeltite), the arsenosulfide CoAsS (cobaltine), and the sulfide Co_3S4 (linneite) (IARC, 1991; WHO, 2006). Cobalt is even found in water in dissolved or ionic form, typically in small amounts (ATDSR, 2004).
Sources of environmental cobalt are both natural and anthropogenic. Natural sources include erosion (wind-blown continental dusts), weathering of rocks and soil, volcanoes, forest fires, extraction by plants, and continental and marine biogenic emissions. Cobalt compounds have been found to occur naturally in seawater, surface water, spring water, and groundwater The major anthropogenic sources of environmental cobalt include mining and processing (smelting) of cobalt-bearing ores, the disposal of cobalt containing waste, and atmospheric deposition from activities such as the burning of fossil fuels and smelting and refining of metals (WHO, 2006). Cobalt metal is usually mixed with other metals to form alloys, that are harder or more resistant to wear and corrosion. These alloys are used in a number of military and industrial applications such as aircraft engines, magnets, and grinding and cutting tools. Cobalt compounds are used as colorants in glass, ceramics, and paints, as catalysts, and as paint driers; they are also used as trace element additives in agriculture and medicine (e.g. artificial hip and knee joints).
^{58}Co and ^{60}Co may be released to the environment as a result of nuclear accidents (i.e, Chernobyl), radioactive waste dumping in the sea or from radioactive waste landfills, and nuclear power plant operations (ATDSR, 2004).
^{60}Co is used for sterilizing medical equipment and consumer products, radiation therapy for treating cancer patients, manufacturing plastics, and irradiating food. ^{57}Co is used in medical and scientific research.
Cobalt and inorganic cobalt compounds are nonvolatile and released into the atmosphere in particulate form. Cobalt released into the atmosphere is deposited on soil, and cobalt

released to water may sorb to particles and settle into sediment or sorb directly to sediment. Cobalt is widely dispersed in the environment in low concentrations. Measured atmospheric concentrations of cobalt are about 1 ng/m^3 or less in non-source areas and generally less than 10 ng/m^3 in source areas, although higher concentrations in source areas have been reported. Cobalt concentrations in drinking water are generally <1–2 µg/l (WHO, 2006).
You may be exposed to small amounts of cobalt by breathing air, drinking water, and eating food containing it. Food and drinking water are the largest sources of exposure to cobalt for the general population. Green vegetables and fresh cereals are the richest sources of cobalt (0.2–0.6 µg/g dry mass) (IARC, 1991; Cobalt Development Institute, 2003).

17.2 EFFECT ON HUMAN HEALTH

Cobalt has both beneficial and harmful effects on human health. As an integral part of the vitamin B12 complex, cobalt is essential in trace amounts for humans and animals, such as cattle and sheep. Exposure of humans and animals to levels of cobalt normally found in the environment is not harmful. However, at higher concentrations, cobalt is toxic to humans and to terrestrial and aquatic animals and plants (ATSDR, 2004; Ministry of Environmental British, 2004).
Exposure to high levels of cobalt can result in lung and heart problems and dermatitis. Liver and kidney have also been affected in animals exposed to high levels of cobalt. Exposure to large amounts of radiations from radioactive cobalt can damage cells in your body. Exposure through food or water to non radioactive cobalt has not been found to cause cancer in humans or in animals. Cancer has been shown, however, in animals who breathed cobalt or when cobalt was placed directly into the muscle or under the skin. Based on the animal data, the International Agency for Research on Cancer (IARC) has determined that cobalt is possibly carcinogenic to humans (ATDSR, 2004).

References:

- Ministry of environmental British Columbia (2004), P.Ag. Water Protection Section Environmental Quality Branch Ministry of Water, Land and Air Protection, Water Quality Ambient Water Quality Guidelines for Cobalt Overview Report, Updated: September 2004.
- ATSDR. (2004). Toxicological profile for cobalt. Atlanta, GA, United States Department of Health and Human Services, Public Health Service, Agency for Toxic Substances and Disease Registry.
- WHO (World Health Organization). (2006). Cobalt And Inorganic Cobalt Compounds, Geneva.
- IARC. (1991). Chlorinated drinking-water; chlorination byproducts; some other halogenated compounds; cobalt and cobalt compounds. Lyon, International Agency for Research on Cancer (IARC Monographs on the Evaluation of Carcinogenic Risks to Humans, Vol. 52).
- Cobalt Development Institute (2003). Cobalt occurrence, supply and demand. Guildford, Surrey, Cobalt Development Institute.

Chapter 18
Copper (Cu)

Cristina Mauceri, Sanjay Mishra.

18.1 ENVIRONMENTAL EFFECT

Copper, atomic number 29, is a reddish transition metal that is stable in its metallic state and forms monovalent (cuprous) and divalent (cupric) cations. It occurs naturally in rock, soil, water, sediment, and, at low levels, in air. Its average concentration in the Earth crust is about 50 parts copper per million parts soil (ppm). It is an essential element for all known living organisms including humans and other animals at low levels of intake. Toxic effects can occur at much higher levels.

Metallic copper is malleable, ductile and a good thermal and electrical conductor. It has many commercial uses because of its versatility. It is used to make electrical wiring, pipes, valves, fittings, coins, cooking utensils and building materials and it is present in munitions, alloys (brass, bronze) and coatings. Copper compounds are used in fungicides, algicides, insecticides and wood preservatives and in electroplating, engraving, lithography, petroleum refining and pyrotechnics; they can be added to fertilizers and animal feeds as a nutrient to support plant and animal growth (ATSDR, 2002) thus they are also used as food additives (e.g., nutrient and/or colouring agent).

Copper enters the environment through releases from copper and other metals mining, and from factories that make or use copper metal or copper compounds, from waste dumps, domestic waste water, combustion of fossil fuels and wastes, wood production, phosphate fertilizer production, and natural sources. Copper is present in the atmosphere and is found in surface water, groundwater, seawater and drinking-water, but it is primarily present in complexes or as particulate matter (ATSDR, 2002). Copper concentrations in surface waters ranged from 0.0005 to 1 mg/l; the median value was 0.01 mg/l (ATSDR, 2002). Copper concentrations in drinking-water vary widely as a result of variations in water characteristics, such as pH, hardness and copper availability in the distribution system.

Human exposure to copper can occur by breathing air, drinking water, eating food, and by skin contact with soil, water and other copper-containing substances. It is excreted from the body in bile, faeces, sweat, hair, menses and urine, over several days.

Food and water are the primary sources of copper exposure in developed countries. Liver and other organ meats, seafood, nuts and seeds (including whole grains) are good sources of dietary copper (IOM, 2001). In general, dietary copper intakes for adults range from 1 to 3 mg/day (IOM, 2001); using a vitamin/mineral supplement will increase exposure by about 2 mg/day. Drinking-water contributes 0.1–1 mg/day in most situations. Thus, daily copper intakes for adults usually range from 1 to 5 mg/day.

The guideline value of WHO for copper in drinking-water is 2 mg/l (WHO, 2011).

18.2 EFFECT ON HUMAN HEALTH

Copper is essential for good health. Copper is readily absorbed from the stomach and small intestine. After nutritional requirements are met, there are several mechanisms that prevent copper overload. Although copper homeostasis plays an important role in the prevention of copper toxicity, exposure to excessive levels of copper can result in a number of adverse health effects including liver and kidney damage, anemia, immunotoxicity, and developmental toxicity. Many of these effects are consistent with oxidative damage to membranes or macromolecules. One of the most commonly reported adverse health effect of copper is gastrointestinal distress. Numerous studies and case reports have reported nausea, vomiting, and/or abdominal pain in humans immediately following ingestion of copper-contaminated water or other beverages stored in a copper or untinned brass container (Araya et al. 2001, 2003a, 2003b, 2003c; Gotteland et al. 2001; Olivares et al. 2001; Pizarro et al., 2001). The observed effects are not usually persistent and gastrointestinal effects have not been linked with other health effects. Individuals ingesting large doses of copper present with gastrointestinal bleeding, haematuria, intravascular haemolysis, methaemoglobinaemia, hepatocellular toxicity, acute renal failure and oliguria. Adults are more resistant than children to copper deficiency symptoms. For adults without genetic disorders of copper homeostasis, long-term intake of copper in the diet at concentrations of 1–10 mg/day has no apparent adverse effects (IOM, 2001). Long-term daily intakes of copper below recommended requirements can lead to anaemia, neutropenia and bone demineralization in malnourished children (IOM, 2001). Copper intakes in the 1–10 mg/day range may have adverse effects for individuals that carry the Wilson disease gene and for infants or children genetically predisposed for one of the childhood copper cirrhosis syndromes (US NRC, 2000). High levels of dietary zinc have a negative effect on copper absorption. High doses of zinc (40-50 mg/day) have been used successfully to treat patients with Wilson disease Zinc limits the amount of copper absorbed (possibly by increasing intestinal metallothionein concentrations and, therefore, slowing the progression of the disease However, high intakes of zinc should be viewed with some concern since copper deficiency may be induced. Conversely, copper supplementation may interfere with zinc absorption (WHO, 1998).

Coughing, sneezing, runny nose, pulmonary fibrosis, and increased vascularity of the nasal mucosa have been reported in workers exposed to copper dust.

Liver damage (necrosis, fibrosis, abnormal biomarkers of liver damage) have been reported in individuals ingesting lethal doses of copper sulfate. The acute lethal dose for adults lies between 4 and 400 mg of copper (II) ion per kg of body weight (WHO, 2011). Liver effects have also been observed in individuals diagnosed with Wilson's disease, Indian childhood cirrhosis, or idiopathic copper toxicosis (that includes Tyrollean infantile cirrhosis). These syndromes are genetic disorders that result in an accumulation of copper in the liver.

The carcinogenicity of copper has not been adequately studied. An increase in cancer risk has been found among copper smelters; however, the increased risk has been attributed to concomitant exposure to arsenic. Increased lung and stomach cancer risks have also been found in copper miners. However, a high occurrence of smoking and exposure to radioactivity, silica, iron, and arsenic obscure the association of copper exposure with carcinogenesis.

Copper (Cu)

The IARC has classified the pesticide, copper 8-hydroxyquinoline, in Group 3, unclassifiable as to carcinogenicity in humans and EPA has classified copper in Group D, not classifiable as to human carcinogenic (IARC, 2002).

References:

- Agency for Toxic Substances and Disease Registry (ATSDR). (2002). Toxicological profile for copper (draft for public comment). Atlanta, GA, US Department of Health and Human Services, Public Health Service, Agency for Toxic Substances and Disease Registry.
- Araya M., McGoldrick M.C., Klevay L.M., Strain J.J., Robson P., Nielsen F., Olivares M., Pizarro F., Johnson L.A., Poirier K.A. (2001). Determination of an acute no-observed-adverse-effect level (NOAEL) for copper in water. *Regul Toxicol Pharmacol* **34**(2):137-48.
- Araya M., Chen B., Klevay L.M., Strain J.J., Johnson L., Robson P., Shi W., Nielsen F., Zhu H., Olivares M., Pizarro F., Haber L.T. (2003a). Confirmation of an acute no-observed-adverse-effect and low-observed-adverse-effect level for copper in bottled drinking water in a multi-site international study. *Reg Tox Pharmacol*, **38**(3):389-99.
- Araya M., Olivares M., Pizarro F., González M., Speisky H., Uauy R. (2003b). Gastrointestinal symptoms and blood indicators of copper load in apparently healthy adults undergoing controlled copper exposure. *Am J Clin Nutr*, **77**(3):646-50.
- Araya M., Pena C., Pizarro F., Olivares M. (2003c). Gastric response to acute copper exposure. *Sci Total Environ*, **303**(3):253-257.
- Gotteland M., Araya M., Pizarro F., Olivares M. (2001). Effect of acute copper exposure on gastrointestinal permeability in healthy volunteers. *Dig Dis Sci*, **46**(9):1909-14.
- IARC. (2002). Overall evaluation of carcinogenicity to humans. Group 3: Unclassifiable as to carcinogenicity to humans. IARC monographs programme on the evaluation of carcinogenic risks to humans. International Agency for Research on Cancer. April 09, 2002. http://193.51.164.11/monoeval/crthgr03htm (accessed on 15 March 2012).
- IOM (Institute of Medicine). (2001). Dietary reference intakes for vitamin A, vitamin K, arsenic, boron, chromium, copper, iodine, iron, manganese, molybdenum, nickel, silicon, vanadium and zinc. Institute of Medicine, Food and Nutrition Board. Washington DC. National Academy Press.
- Olivares M., Araya M., Pizarro F., Uauy R. (2001). Nausea threshold in apparently healthy individuals who drink fluids containing graded concentrations of copper. *Regul Toxicol Pharmacol*, **33**(3):271-75.
- Pizarro F., Olivares M., Araya M., Gidi V., Uauy R. (2001). Gastrointestinal effects associated with soluble and insoluble copper in drinking water. *Environ Health Perspect*, **109**(9):949-52.
- WHO EHC 200, Copper, 1998.
- WHO (World Health Organization). (2011). Guidelines for Drinking-water Quality. 4th ed. WHO, Geneva, Switzerland. ISBN 978 92 4 154815 1.

Chapter 19
Iron (Fe)

Margherita Ferrante

19.1 ENVIRONMENTAL EFFECT

Iron, atomic number 26, is the second most abundant metal in the earth's crust, of which it accounts for about 5%. Elemental iron is rarely found in nature. Iron is used as constructional material. Iron oxides are used as pigments in paints and plastics, but also as food colours and for the treatment of Fe deficiency in humans. Some iron salts are used as coagulants in water treatment. Aeration of Fe-containing layers in the soil can affect the quality of both groundwater and surface water if the groundwater table is lowered or nitrate leaching takes place. Dissolution of iron can occur as a result of oxidation and decrease in pH.

In anaerobic groundwater where iron is in the form of Fe(II), concentrations will usually be 0.5–10 mg/l, but concentrations up to 50 mg/l can be found. Iron concentrations in drinking-water are normally less than 0.3 mg/l and no guideline value for iron in drinking-water is proposed (WHO, 2011).

19.2 EFFECT ON HUMAN HEALTH

Iron is an essential element. It is central to oxido-reductive processes in humans and in oxygen transport. Iron is a constituent of numerous essential proteins such as haem-containing proteins like the cytochromes, Fe-sulphur proteins, Fe-activated enzymes and proteins for iron storage and transport (Arredondo and Nunez 2005; Fraga 2005). Thus, iron is involved in many crucial processes in the body, e.g. oxygen transfer from lungs to tissues and trans-membrane electron transfer. Iron is also important for adequate immune response and neuronal functions, that means appropriate activity and numbers of immune cells (macrophages, neutrophils and T-lymphocyte), and early life myelination (Arredondo and Nunez, 2005).

Food is the main source of iron in the general population. Sources of haem iron (15% of consumption) are from animal tissue.. Sources of non-haem iron are cereals, seeds of leguminous plants, fruits, vegetables, and dairy products (Fraga, 2005). Daily consumption of iron originating from food ranges from 10 to 14 mg (WHO, 2011). Drinking-water containing 0.3 mg/l would contribute about 0.6 mg to the daily intake, about 2% to 8% of the recommended dietary allowance (RDA) for iron.

Cellular iron homeostasis is regulated by the transferrin receptor (Fe transport), and ferritin (Fe-storage protein). It is achieved by iron regulated intestinal absorption. The concept of the "erythropoietic regulator" emerged from the observation that intestinal iron absorption increases when erythropoietic needs for iron are not satisfied. Similarly, the concept of "stores regulator" emerged from the observation that iron absorption responds inversely to the size of iron stores (Arredondo and Nunez, 2005).

The most common problem for humans with iron deficiency is anaemia.
The dietary intake level for iron that is sufficient to meet the nutrient requirements of nearly all healthy individuals in a life stage and gender group, that is called recommended dietary allowance (RDA) range from 7 to 27 mg/day depending on age, gender and physiological status (IOM, 2001). Parous women and children are those most likely to be iron-deficient. However a normal diet generally provides the iron that is needed (Dvorak et al, 2007).
Symptoms associated with iron deficiency are tiredness, lassitude, and general feelings of lack of energy, most of them having their origin in lack of transported oxygen to tissues. In addition, the immune system may be depressed, since iron is essential for proper enzyme function in immune cells and signalling pathways (Bowman and Russell, 2006). Iron deficient infants may suffer from delayed motor maturation and cognitive dysfunction, impaired growth and general weakness, and depressed intestinal function (Ryan, 1997).
Iron deficiency is more common than overload. Children and pregnant women usually suffer from anemia. Acute iron poisoning occurs mostly accidentally, when large numbers of iron containing pills are ingested usually by child. It is followed by symptoms like severe vomiting, diarrhoea, abdominal pain but also it may cause conjunctivitis, choroiditis, and retinitis if it comes into contact and remains in the tissues.
Chronic iron intoxication is occurs frequently associated to genetic and metabolic diseases, repeated blood transfusions, or excessive intake (Fraga and Oteiza, 2002).
A subset of the population suffers from iron overload or *Hemochromatosis.* Wiley & Moore, 2009) Hereditary hemochromatosis (HH) is only one of several iron loading diseases but it is the most common genetic disease but also tragically the most undiagnosed in people of Celtic heritage (ancestors from Ireland, Wales, England). This people are at high risk because of carrying the HFE mutations for hereditary hemochromatosis.
The symptoms of hemochromatosis can include: chronic fatigue, arthritis, heart disease, cirrhosis, cancer, diabetes, thyroid disease, impotence, and sterility (AHS, 2012).
The average lethal dose of iron is 200–250 mg/kg of body weight, but death has occurred following the ingestion of doses as low as 40 mg/kg of body weight (WHO, 2003). High level of iron in drinking water has been identified as a catalyst for oxidative stress,
ss, it stimulates the growth of bacteria, and it may increase the likelihood of autoimmune diseases in genetically predisposed individuals. Iron is associated with oxidative free radicals, and thus may be a factor in aging. Patients with Alzheimer or Parkinson disease frequently have increased brain iron content. Fe may also play a role in atherosclerosis and diabetes (Mitchell et al., 2011).
An intake of 0.4–1 mg/kg of body weight per day is unlikely to cause adverse effects in healthy persons (WHO, 2011). With iron water concentration of 0,3 mg/l, body weight of 60 kg and daily water consumption of 2l, an intake of 0,01mg Fe/kg of body weight comes from drinking water.
The lack of mechanisms for cell secretion and body excretion causes iron to accumulate in cells and the body. In cells this gives rise to the labile or reactive iron pool that is probably involved in the induction of oxidative damage to vital cellular components, whereas body iron accumulation may give rise, in time, to neurodegenerative and aging processes (Arredondo and Nunez, 2005).
Based on provisional maximum tolerable daily intake (PMTDI) of 0.8 mg Fe/kg of body weight given by the Joint Expert Committee on Food Additives (JECFA), the allocation of 10% of this PMTDI to drinking-water gives a value of about 2 mg/l, which does not present a

hazard to health (WHO, 2011). WHO has not yet set health-based guideline value for iron, because at levels found in drinking-water no adverse effects have been found.Recent studies from India (Subba Rao, 2008) suggest that when iron is ingested with drinking water at higher concentrations it may cause serious health problems. The Indian study showed that when the iron concentration in drinking water exceeded 0.400-0.780 mg/l health changes were reported on e.g. the gastrointestinal tract. It was found that people using for drinking purposes such water suffered from severe constipation, aching neck, hips and arms, head ache and gastric ulcer. In addition muscle cramps during nights and he had rapid pulse, at instants were observed. It was also noticed that rusty sweat appeared and white shirts became rusty. Other symptoms were observed such as diarrhoea, lethargy, loss of appetite and weight, apathy.

It should be underlined that iron concentrations in drinking water mentioned in those studies were exceptionally high (up to 25 times higher than EU parametric value). Acute iron poisoning occurs mostly accidentally. There is no evidence of toxicity from drinking water with commonly found iron concentrations.

References:

- AHS. American Hemochromatosis Society. (2012). http://www.americanhs.org/faq.htm (accessed on 5 september 2012).
- Arredondo M., Nu´n˜ez M.T. (2005). Iron and copper metabolism. *Molecular Aspects of Medicine,* **26**:313–27.
- Bowman B.A., Russell R.M. (2006). Present knowledge in nutrition. 9th ed. Volume 1. Washington, D.C.: ILSI Press.
- Dvorak B.I., Prasai G., Skipton S.O., Woldt W. (2007). Drinking Water: Iron and Manganese. In Water Resources Management. Drinking Water. NIANR. University of Nebraska-Lincoln Extention, Institute of Agriculture and Natural Resources. http://www.ianrpubs.unl.edu/epublic/pages/index.jsp (accessed on 5 september 2012).
- Fraga C.G. (2005). Relevance, essentiality and toxicity of trace elements in human health. *Molecular Aspects of Medicine*, **26**:235–44.
- Fraga C.G., Oteiza P.I. (2002). Iron toxicity and antioxidant nutrients. *Toxicology*, **180**:23-32.
- IOM. (2001). Institute of Medicine of the National Academies. Report: Dietary Reference Intakes for Vitamin A, Vitamin K, Arsenic, Boron, Chromium, Copper, Iodine, Iron, Manganese, Molybdenum, Nickel, Silicon and Zinc.
- Mitchell E., Frisbie S., Sarkar B. (2011). Exposure to multiple metals from groundwater a global crisis: Geology, climate change, health effects, testing, and mitigation. *Metallomics*, **3**:874–908.
- Ryan A.S. (1997). Iron-deficiency anemia in infant development: Implications for growth, cognitive development, resistance to infection, and iron supplementation. *American Journal of Physical Anthropology*, **104**(25):25-62.
- Subba Rao N. (2008). Iron content in groundwaters of Visakhapatnam environs, Andhra Pradesh, India. *Environmental Monitoring and Assessment*, **136**:437-47.

- WHO (World Health Organization). (2011). Guidelines for Drinking-water Quality. 4th ed. WHO, Geneva, Switzerland. ISBN 978 92 4 154815 1.
- Wiley J., Moore M.R. (2009). Heme biosynthesis and its Disorders in Hoffmann et al. Hematology Principles & Practice 5th Edition. pp 447-490.

Chapter 20
Lanthanum (La)

Michael R. Moore

20.1 ENVIRONMENTAL EFFECT

Lanthanum, atomic number 57, is the first of the group of 15 elements known as the 'lanthanides', or as the rare earth elements Lanthanum was discovered in 1839 by Carl G. Mosander. Native lanthanum is a mixture of two stable isotopes,138 and 139. Lanthanum is a soft, malleable, ductile, silver-white metal. It is chemically active as one of the most reactive of the rare-earth metals: it oxidizes rapidly in air and it reacts with water to form the hydroxide. It is the most electropositive (cationic) element of the rare earth group, is always trivalent, and its binding is almost exclusively ionic. It is a hard 'acceptor' with a preference for oxygen-containing anions. The most common biological ligands are with carboxyl and phosphate groups with which it forms tight complexes (Persy et al., 2006)

Lanthanum carbonate has been used pharmaceuticallly as a phosphate binding agent for patients with hyperphosphataemia Another application of lanthanum is found to be in water treatment, for removing oxyanions, such as phosphate and arsenate.

Lanthanum (La^{3+}) is not influenced by redox reactions (as in the case of Al $^{3+}$),

And when bound with PO4 forms the insoluble compound, LaPO4. Lanthanum and lanthanum salts are not on the NOHSC List of Designated Hazardous Substances (NOHSC, 1999a) However, it has been found that free Lanthanum (from the dissolution of $LaCl_3$) can be toxic to aquatic organisms such as Daphnia (Peterson et al. ,1974; Barry & Meehan 2000). This limited its use, until an appropriate carrier that could lock the lanthanum ions into its structure was discovered in the mid 1990s by CSIRO in Australia. Lanthanum ecotoxicity was greatly reduced by incorporating the lanthanum ions into thematrix of the mineral Bentonite.

20.2 EFFECT ON HUMANS.

Lanthanum toxicity to humans has been studied because of its therapeutic use as lanthanum carbonate. Phosphate accumulation in the human body and hyperphosphatemia are associated with an increased risk of death. Lanthanum carbonate an oral phosphate binder is used for treatment of hyperphosphatemia in patients with chronic kidney disease who are undergoing dialysis. Lanthanum carbonate dissociates in the acid environment of the upper gastrointestinal tract to release lanthanum ions that allow the formation of lanthanum phosphate. This insoluble lanthanum phosphate is eliminated in the faeces without significant absorption of lanthanum. Unlike Al there is little accumulation of lanthanum in the body of dialysis patients, primarily because of its very low gastrointestinal absorption and elimination in faeces. The kidney is not significantly involved in the clearance of lanthanum; the main excretion route for absorbed lanthanum being via the liver into bile (Damment & Pennick, 2007; Das et al. 1988). Clinical studies of up to four years have not found any hepatotoxic effect of lanthanum in patients treated with it (Persy et al. 2006). Patients treated with lanthanum carbonate for one year did not experience any of the aluminium-like toxic effects on bone such as osteomalacia (De Bror & D'Haese, 2004). No effects of lanthanum on bone have been observed in animals with normal renal function loaded with lanthanum at doses up to 2000 mg/kg/day for two years (Damment et al., 2003). On the other hand, rats with chronic renal failure loaded with very high doses of lanthanum carbonate for 12 weeks showed an impairment of bone mineralisation (Behets et al., 2004). However, further studies showed that tthese effects were a consequence of phosphate depletion induced by the administration of high doses of lanthanum carbonate which could be reversed by administration of phosphate (Persy et al., 2006; Damment & Shen, 2005). However, the effects of lanthanides on the central nervous system, are still unclear. Long-term lanthanum exposure in rats results in persistent alternations in nervous system function which were associated with changes in the distributions of elements such as Ca, Fe and Zn in the brain (Feng et al., 2006).

References:

- Barry M.J., Meehan B.J. (2000). The acute and chronic toxicity of lanthanum to Daphnia carinata. Chemosphere, **1**(10):1669-74
- Behets G.J., Verberckmoes S. C., D'Haese P. C., de Broe M . E. (2004). Lanthanum carbonate: a new phosphate binder. Curr Opin Nephrol Hypertens, **13**: 403 –409.
- Behets G.J., Verberckmoes S.C., Oste L., Bervoets A.R., Salome M., Cox A.G.,Denton J., De Broe M.E,and D'Haese P.C. (2005). Localisation of lanthanum in bone of chronic renal failure rats after oral dosing with lanthanum carbonate. Kidney Int, **67**:1830–1836.
- Damment S.J.P., Gill M. (2003). The pharmacokinetics and tissue distribution of Lanthanum carbonate (Fosrenol®), a new non-aluminum, non-calcium phosphate binder(abstract). J Am Soc Nephrol, **14**: 204A.
- Damment S.J.P., Shen V. (2005). Assessment of effects of lanthanum carbonate with and without phosphate supplementationon bone mineralization in uremic rats. Clin Nephrol, **63**: 127–137.
- Damment S.J.P., Pennick M. (2007). Systemic lanthanum is excreted in the bile of rats. Toxicol Lett, **171**:69–77.
- Das T., Sharma A., Talukder G. (1988). Effects of lanthanum in cellular systems. Biological Trace Element Research, **18**:201– 228.
- De Broe M.E., D'Haese P.C. (2004). Improving outcomes in hyperphosphataemia. Nephrol Dial Transplant, **19** (Suppl 1): i14–i18.
- Feng L., Xiao H., He X., Li Z., Li F., Liu N., Zhao Y., Huang Y., Zhang Z., Chai Z. (2006). Neurotoxicological consequence of long-term exposure to lanthanum.Toxicol Lett.; **165**(2):112-20.
- National Occupational Health and Safety Commission (NOHSC). 1999a. List of Designated Hazardous Substances [NOHSC:10005(1999)]. Australian Government Publishing Service, Canberra, Australia.

Lanthanum (La)

- Persy V.P., Behets G.J, Bervoets A.R., De Broe M.E., D'Haese P.C. (2006). Lanthanum: A safe phosphate binder. Seminars in Dialysis, **19**(3): 195–199.
- Peterson S.A., Sanville W.D., Stay F.S., Powers C.F. (1974). Nutrient Inactivation as a lake Restoration Procedure-Laboratory Investigations. National Environmental Research Center. US EPA, Carvallis, Oregon. Report number: EPA-660/3-74-032.

Chapter 21
Lead (Pb)

Zorica Rasic-Milutinovic, Dragana Jovanovic, Ignatius Maduka.

21.1 ENVIRONMENTAL EFFECT

Lead (Pb) is usually found in combination with other elements to form different minerals. The mineral called galena is most often mined to produce lead metal. There are so many sources of lead in our environment that can be poisonous for humans. The major sources of lead exposure include lead in paint, gasoline, water distribution systems, food, and lead used in hobby activities. Lead exposure attributable to automobile air emissions was a major exposure source prior to 1976. Between 1976 and 1990, lead used in gasoline declined by 99.8% in the United States, but not in some other countries where lead is permitted in gasoline: National Association of Physicians for the Environment (NAPE, 1993). Lead poisoning is an environmental and public health hazard of global proportions. Children and adults in virtually every region of the world are being exposed to unsafe levels of lead in the environment. In fact, children are exposed to lead from different sources, such as paint, gasoline, and solder, and through different pathways such as air, food, water, dust and soil. It can be from a single high level exposure or the cumulative effect of repeated high or low level exposures. Once introduced into an environment, Lead can remain in a dangerous form for a very long time.

In the U.S., the major sources of lead exposure culpability have shifted from gasoline to paint and water distribution systems and fixtures. On the other hand, in Eastern Europe and many developing countries, the air pollution contribution to lead exposure remains high.

The primary source of lead in drinking water is from lead pipes, brass fittings, soldered joints, and other service and dwellings connections, used to distribute drinking water to consumers (IWA, 2010).

The guideline value of WHO for lead in drinking-water is 0.01 mg/l (WHO, 2011).

21.2 EFFECT ON HUMAN HEALTH

Children are more susceptible to lead than adult due to higher gastrointestinal uptake and the permeable blood–brain barrier (Jarup, 2003). The negative effects of exposure to lead are various (see Tab. 21.1), including behavioural disturbances, learning and concentration difficulties and diminished intellectual capacity. For every 0.48 μmol/l (10 μg/dl) increase in blood lead level a weighted mean IQ is decreased for 2 points (WHO, 1995; WHO, 2011). Lead inhibits several enzymes in heme formation pathway, including delta-ALAD, coproporphyrinogen oxidase and ferrochelates. Because lead effectively inhibits ALAD activity, resulting in accumulation of ALA in blood and urine, urinary ALA has also been used as a biomarker for lead exposure or a marker of early biologic effect of lead. ALA seems to be a contribution factor to lead-induced neurotoxicity. The pathogenetic effect of lead is multifactorial since it directly interrupts the activity of enzymes, competitively inhibits absorption of important trace minerals and deactivates antioxidant sulphydryl pools (Jomova and Valko, 2011). Besides its ability to induce brain disorders, lead may also cause hypertension, kidney damage, anemia and adverse reproductive outcomes (Kakkar and Jaffery, 2004).

Lead compounds do not appear to cause genetic damage directly, but may do so through several indirect mechanisms, including inhibition of DNA synthesis and repair, oxidative damage, and interaction with DNA-binding proteins and tumor-suppressor proteins (NTP, 2011). Significant increase in frequency of chromosomal aberrations with increasing years of lead exposure suggested lead as the causal factor for genotoxicity. In a population exposed to lead there was significant increase in mean chromosomal aberrations, micronuclei and sister chromatid exchanges (Vij, 2009). Inorganic lead compounds are classified as probably carcinogenic to humans (group 2A) by IARC (IARC, 2004).

Table 21.1 Toxic effect of lead (WHO, 2001)
Hyperactivity
Retard of cognitive development
Learning disabilities
Slow growth
Loss of synaptic plasticity
Headaches
Hearing loss
Neuropathy
Damage of Kidney functionality
Loss of dermal sensitivity
Damage on heme formation pathway
Genotoxicity

References:

- International Agency for the Research on Cancer (IARC). (2004). Inorganic and organic lead compounds. IARC monographs on the evaluation of carcinogenic risks to humans. Vol 87. Lyon.
- International Water Assocciation (IWA). (2010). Best Practice Guide on the Control of Lead in Drinking Water. IWA Publishing; ISBN 9781843393696.
- Kakkar P., Jaffery F.N. (2005). Biological markers for metal toxicity. *Environmental Toxicology and Pharmacology*; **19**:335–49.
- Jarup L. (2003). Hazards of heavy metal contamination. *British Medical Bulletin*; **68**:167–82.
- Jomova K., Valko M. (2011). Advances in metal-induced oxidative stress and human disease. *Toxicol*; **283**:65–87.

- National Association of Physicians for Environment (NAPE). Conference Summary. http.//www.intr.net/napenet/airsum.html Thron, 1993, (accessed on 1 March 2012).tional Toxicology Program (NTP). Report on Carcinogens. National Toxicology Program, 12th ed., 2011, http://ntp.niehs.nih.gov/ntp/roc/twelfth/roc12.pdf (accessed on 10 March 2012).
- Vij A.G. (2009). Hemopoietic, Hemostatic and Mutagenic Effects of Lead and Possible Prevention by Zinc and Vitamin C. *Al Ameen J Med Sc*, **2**(2):27-36.
- WHO (World Health Organization). (1995*). Lead. Environmental Health Criteria*, vol. 165. Geneva, Swizterland.
- WHO (World Health Organization). (2011). Guidelines for Drinking-water Quality. 4th ed. WHO, Geneva, Switzerland. ISBN 978 92 4 154815 1.

Chapter 22
Lithium (Li)

Adriana Floridia, Maria Grazia D'Agati.

22.1 ENVIRONMENTAL EFFECT

Lithium, atomic number 3, the lightest of the alkali metals, was discovered in 1817 by Johan August Arfvedson in petalite, a tektosilicate of composition $LiAlSi_4Si_{10}$. The element occurs in numerous other minerals and was named after the Greek *lithos*, stone, because of its presence, in trace amounts, in virtually all rocks (Gmelin's Handbook, 1960). Mobilized by weathering processes, lithium is transported into soils, from which it is taken up by plants and enters the food chain. Falls under water hazard class 1, weakly harmful in water. Lithium is not a very big threat to flora and fauna, nor in terrestrial or in aquatic environments. It is readily absorbed by plants, causing plants to be an indicator of soil lithium concentrations. Lithium is not a dietary mineral for plants, but it does stimulate plant growth. Too much lithium may be toxic. To prevent toxicity, calcium may be added to soils to prevent uptake of lighter minerals. The amount of lithium in plants usually lies between 0.2 and 30 ppm.

Lithium was detected in human organs and fetal tissues in the late 19th century, leading to early suggestions of possible specific functions in the organism (Scha¨fer, 2000), but medical applications of lithium preceded studies on lithium as an essential micronutrient From the mid 1800s to the early 1940s, lithium carbonate was used - without adequate substantiation of success - to treat gout and to dissolve urate bladder stones. The first legitimate medical application of lithium was introduced 1949, when lithium carbonate was found to be beneficial in manic depressive illness (Cade, 1949).

Today, lithium carbonate is one of the most widely prescribed psychiatric drugs. More recently, it has found other applications, notably in oncology (Kehrberg, 1991; Hager et al., 2000) and in dermatology (Horrobin, 1991).

Natural lithium traces in water were considered by John Cade in 1949 to have the potential to influence mental health. Meanwhile, the mood-stabilising effects of lithium are widely recognised by the psychiatric community (Altamura et al., 2011) and its suicide preventive properties are well documented (Mu¨ller-Oerlinghausen & Lewitzka, 2010). As a natural trace element, lithium is mobilised by rain from rock and soil and dissolves in ground and drinking water. In some geographic regions, its concentrations may reach up to 5.2 mg/l, reflecting a natural daily intake of lithium of up to 10 mg/day. Lithium concentrations in drinking water vary considerably by geographic region, (Schrauzer, 2002; Concha et al., 2010) and correlate with natural lithium resources.

Seawater contains approximately 0.17 ppm lithium. Rivers generally contain only 3 ppb, whereas mineral water contains 0.05-1mg lithium per liter. In solution it is only found as Li^+ (aq).

22.2 EFFECT ON HUMAN HEALTH

The amount of lithium in the human body is approximately 7 mg. Although lithium is not an essential element, it may influence metabolism and animal's physiology (see Tab. 22.1). Lithium is absorbed via sodium channels in the small intestine and uniformly distributed in body water, although others have found differences in lithium levels between tissues (Schrauzer, 2002) and plasma and brain concentrations (Soares et al., 2001).

Table 22.1 Physiological functions of lithium in the human body
Maintenance of the physiological balance of neurotransmitter (Schrauzer, 2002)
Maintenance of mood (Hughes et al., 2000)
Regulation of gene expression by transcriptional factor (Roberts et al., 2007)
Prevention Alzheimer's disease (Forlenza et al., 2012)

Because renal clearance is not dependent on plasma lithium levels, plasma levels are proportional to daily intake. Although excreted mainly by the kidney, approximately 80% of lithium is reabsorbed by the proximal renal tubule. Excretion of lithium is dependent on the glomerular filtration rate and therefore affected by renal diseases and age, conditions in which plasma lithium increases. On the other hand, dehydration (and loss of salt) decreases the clearance of lithium (Soares et al., 2001). These considerations suggest that water intake has at least a twofold effect on plasma lithium levels, and that lithium retention is probable when intake of water is reduced.

It has to be noted that lithium concentrations increase in the brain during the first trimester of gestation (Schrauzer, 2002) and early exposure to lithium may cause damage in human brain neurodevelopment (Gentile, 2010). Lithium acts on mood and suicidality via complex interactions with the serotoninergic system (Hughes et al., 2000) and most recent studies suggest that lithium has stimulating effects on neurogenesis (Wada, 2009), which could explain both toxicity during neurodevelopment as well as antidepressive/ antisuicidal effects, Currently, not enough is known about the effects of lithium on the prevalence of neurodevelopmental disorders to consider artificially increasing its levels in drinking water as a method of universal prevention.

There is evidence that lithium concentrations in drinking water are inversely correlated with suicide rates. Starting with anecdotic reports about the beneficial effects of lithium in drinking water on mental health in 1949 and earlier, there is increasing evidence from three independent countries and continents that lithium in drinking water is associated with reduced mortality from suicide (Nestor et al., 2011). Consistent with this Dawson et al. (1972) linked low lithium intakes with altered behavior and aggressiveness in humans.

Although it can be assumed that lithium in drinking water explains a part of the variance in suicide mortality, other possible lithium sources should be mentioned. It has also been considered that lithium could play a role during the cooking process (Krachler & Shotyk, 2009). Indeed, lithium as a salt is likely to be taken up from drinking water into vegetables and animal-derived food and vice versa during osmotic processes. Lithium levels in food would then regress to the levels of local cooking/drinking water. Moreover it needs to be considered whether the large water volumes used during bathing and showering could be an additional source of natural lithium (Concha et al., 2010).

The side-effects of chronic lithium medication are well documented and include, gastrointestinal pain, diarrhoea, tremor, polyuria, weight gain, edema, and hypothyroidism, as well as renal tubular damage, resulting in reduced urinary concentrating capacity, and decreased glomerular filtration rate (Aral & Vecchio-Sadus, 2008; Grandjean & Aubry, 2009). However, little is known about the effects of long-term exposure to elevated lithium levels through drinking water (Ohgami et al., 2009). Some studies showed that lithium passes to the fetus and breast milk, hence, elevated concentrations in drinking water implies considerable exposure very early in life. Moreover studies indicate that medication during pregnancy increases the risk of developmental toxicity, in particular kidney toxicity, nephrogenic diabetes insipidus, euthyroid goiter, and hypoglycemia (Yoshioka et al., 2009; Schou, 1998), which cause concern for developmental effects also because of the markedly high concentrations in drinking water (Concha et al., 2010).

As Li deficiency in humans is unlikely ever to reach the degree of severity observed in experimental Li-depleted animals, any symptoms of lithium deficiency in humans, if at all observable, would be expected to be mild and manifest themselves primarily by behavioral rather than physiological abnormalities.

References:

- Altamura A.C., Lietti L., Dobrea C., Benatti B., Arici C., Dell'osso B. (2011). Mood stabilizers for patients with bipolar disorder: the state of the art. *Expert Rev Neurother*; **11**: 85–99.
- Aral H.; Vecchio-Sadus A. (2008). Toxicity of lithium to humans and the environment--a literature review. *Ecotoxicol. Environ. Saf.* **70**(3):349–56.
- Cade J.F.J. (1949). Lithium salts in the treatment of psychotic excitement. *Med J Aust*, 349–52.
- Concha G., Broberg K., Grande´r M., Cardozo A., Palm B., Vahter M. (2010). High-level exposure to lithium, boron, cesium, and arsenic via drinking water in the Andes of northern Argentina. *Environ Sci Technol*; **44**:6875–80.
- Dawson E.P., Moore T.D., McGanity W.J. (1970). The mathematical relationship of drinking water lithium and rainfall on mental hospital admission. *Dis Nerv Syst*, **31**:1–10.
- Dawson E.P., Moore T.D., McGanity W.J. (1972). Relationship of lithium metabolism to mental hospital admission and homicide. *Dis Nerv Syst*, **33**:546–556.
- Dawson E.B. (1991). The relationship of tap water and physiological levels of lithium to mental hospital admission and homicide in Texas. In Schrauzer GN, KlippelKF (eds): Lithium in Biology and Medicine. Weinheim: VCH Verlag, pp 171–87.

- Gentile S. (2010). Neurodevelopmental effects of prenatal exposure to psychotropic medications. *Depress Anxiety*; **27**:675–86.
- Gmelin's Handbook of Inorganic Chemistry. (1960). 8th ed, System No. 20. Weinheim: Verlag Chemie.
- Grandjean E.M., Aubry J.M. (2009). Lithium: updated human knowledge using an evidence-based approach. Part III: clinical safety. *CNS Drugs*, **23**(5):397–418.
- Hager E.D., Dziambor H., Winkler P., Macholdt K. (2000). Thrombopoietischer Effekt von Lithiumcarbonat bei Patienten mit niedriger Thrombozytenzahl nach einer Chemo- oder Radiotherapie. In Anke M et al. (eds): Proceedings Mengen und Spurenelemente, 20. Jahrestagung. Leipzig: Schubert-Verlag, pp 119–29.
- Horrobin D.F. (1991). Lithium effects on fatty acid metabolism and their role in therapy of seborrhoeic dermatitis and herpes infections. In Schrauzer GN, Klippel KF: Lithium in Biology and Medicine. Weinheim: VCH Verlag, pp 67–72.
- Hughes J.H., Dunne F., Young A.H. (2000). Effects of acute tryptophan depletion on mood and suicidal ideation in bipolar patients symptomatically stable on lithium. *Br J Psychiatry*; **177**:447–51.
- Yoshioka W., Akagi T., Nishimura N., Shimizu H., Watanabe C., Tohyama C. (2009). Severe toxicity and cyclooxygenase (COX)-2 mRNA increase by lithium in the neonatal mouse kidney. *J. Toxicol. Sci*, **34**(5):519–52.
- Kapusta N.D., Mossaheb N., Etzersdorfer E., Hlavin G., Thau K., Praschak-Rieder M.W.N., Sonneck G., Leithner-Dziubas K. (2011). Lithium in drinking water and suicide mortality. *The British Journal of Psychiatry*, **198**:346–50.
- Kehrberg G. (1991). Study of the prophylactic effect of lithium in radiogenic leucocytopenia. In Schrauzer GN, Klippel K-F (eds): Lithium in Biology and Medicine. Weinheim: VCH Verlag, pp 49–63.
- Krachler M., Shotyk W. (2009). Trace and ultratrace metals in bottled waters: survey of sources worldwide and comparison with refillable metal bottles. *Sci Total Environ*; **407**:1089–96.
- Mu¨ller-Oerlinghausen B., Lewitzka U. (2010). Lithium reduces pathological aggression and suicidality: a mini-review. *Neuropsychobiology*; **62**:43–9.
- Ohgami H., Terao T., Shiotsuki I., Ishii N., Iwata N. (2009). Lithium levels in drinking water and risk of suicide. *Br. J. Psychiatry*, **194**(5):464–65.
- Roberts J, Scott AC, Howard MR, Breen G, Bubb VJ, Klenova E, Quinn JP. (2007). Differential regulation of the serotonin transporter gene by lithium is mediated by transcription factors, CCCTC binding protein and Y-box binding protein 1, through the polymorphic intron 2 variable number tandem repeat. *J Neurosci.* **14**;27(11):2793-801.
- Scha¨fer U. (2000). The development of lithium from chemical laboratory curiosity to an ultratrace element, a potent drug and a versatile industrial material. In Seifert, M, Langer U, Scha¨fer U, Anke M (eds): Mengen und Spurenelemente, Author and Element Index 1981-2000. Leipzig: Schubert-Verlag, pp 21–9.
- Schou M. (1998). Treating recurrent affective disorders during and after pregnancy. What can be taken safely? *Drug Saf*, **18**(2):143–52.
- Schrauzer G.N. (2002). Lithium: occurrence, dietary intakes, nutritional essentiality. *J Am Coll Nutr*, **21**:14–21.

- Soares J.C., Boada F., Spencer S., Mallinger A.G., Dippold C.S., Wells K.F., Frank E., Keshavan M.S., Gershon S., Kupfer D.J. (2001). Brain lithium concentrations in bipolar sorder patients: preliminary ($_7$)Li magnetic resonance studies at 3 T. *Biol Psychiatry*; **49**:437–43.
- Wada A. (2009). Lithium and neuropsychiatric therapeutics: neuroplasticity via glycogen synthase kinase-3beta, beta-catenin, and neurotrophin cascades. *J Pharmacol Sci*; **110**:14–2.

Chapter 23
Magnesium (Mg)

Caterina Ledda, Maria Fiore

23.1 ENVIRONMENTAL EFFECT

Magnesium, Atomic Number 12, is an element with symbol Mg. It is an alkaline earth metal and the eighth most abundant element in the Earth's crust and ninth in the known universe as a whole (Housecroft & Sharpe, 2008).
Due to magnesium ion's high solubility in water, it is the third most abundant element dissolved in seawater (Turekian, 1968).
In vegetation magnesium is the metallic ion of chlorophyll, and for thisreason is a common additive to fertilizers.

23.2 EFFECT ON HUMAN HEALTH

Magnesium is the eleventh most abundant element by mass in the human body; its ions are essential to all living cells, where they play a major role in important macromolecules as a ATP, DNA, and RNA. Magnesium is very important for enzyme function (Fig. 23.1). Magnesium compounds are used medicinally as common laxatives, antacids (e.g. magnesia), and in a number of situations where stabilization of abnormal nerve excitation and blood vessel spasm is required. Magnesium ions in low concentrations help to impart a natural tartness to mineral waters.

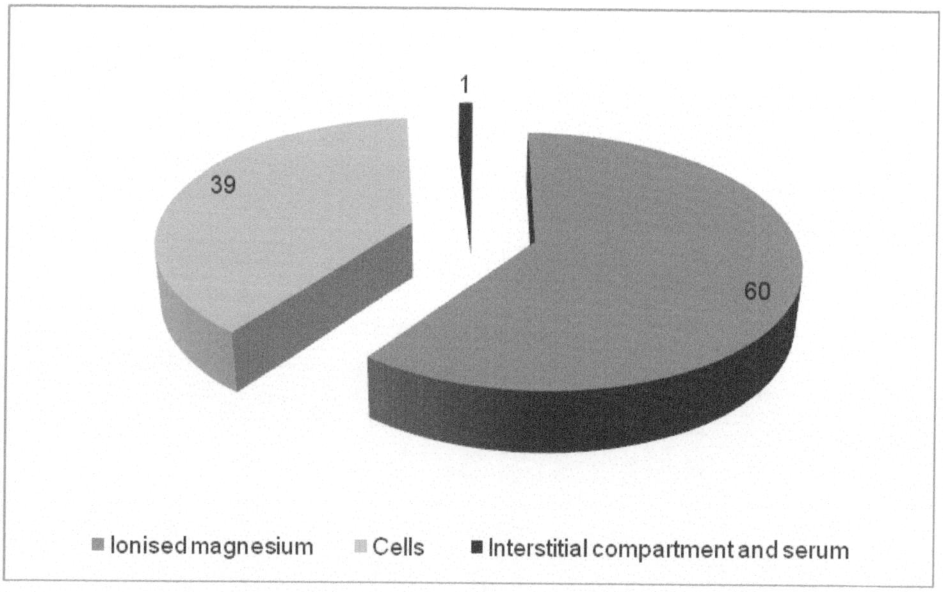

Fig. 23.1: Mg content (%) in human body

Magnesium is a cofactor for approximately 300 enzymes and it is important for the carbohydrate metabolism, heart, muscles, bones and nerve impulses, therefore it has an important role in the proper functioning of heart and blood vessels, and is also helpful in diabetes and for improved longevity. Magnesium is an essential component of a healthy human diet (see Table 23.1).

Table 23.1 Toxic effect of lead (WHO, 2001)
Hyperactivity
Retard of cognitive development
Learning disabilities
Slow growth
Loss of synaptic plasticity
Headaches
Hearing loss
Neuropathy
Damage of Kidney functionality
Loss of dermal sensitivity
Damage on heme formation pathway
Genotoxicity

Human magnesium deficiency is rare, low levels of magnesium in the body has been associated, with the development of a number of human illnesses such as asthma, diabetes, and osteoporosis (Ciarallo et al., 2000; Bo & Pisu, 2008; Bartlett & Eperjesi, 2008; Fallico et al., 1984). Magnesium plays a role in preventing stroke and heart attack (Champagne, 2008; Chiuve et al., 2011).

RDI for Magnesium is about 300 - 420 mg (Table 23.1). Recently some experts have suggested higher RDI value, of 450-500 mg, for preventing Coronary Heart Disease (Chiuve et al., 2011).

Such higher intake may be difficult to be achieved from normal food unless magnesium rich drinking water is used. You may have muscle weakness, confusion, and decreased reflexes if your blood test results show severely low magnesium levels (WHO, 2004).

Magnesium is crucial for cellular energy metabolism, protein and nucleic acid synthesis, normal vascular tone and insulin sensitivity. Heart, circulation and nervous system have also magnesium dependant functions in our body. The importance of magnesium intake from drinking water was stated in a Taiwan study, as the risk of rectal cancer from THMs (Trihalomethanes) was increased when the magnesium level was low in drinking water (Kuo et al., 2010). Drinking water containing up to 100 mg/L magnesium is to be regarded safe, but levels of 10-50 mg/L is recommendable for healthy individuals. However, those with kidney failure will not be able to tolerate magnesium rich water (WHO, 2004).

References:

- Bartlett H.E., Eperjesi F. (2008). Nutritional supplementation for type 2 diabetes: a systematic review. *Ophthalmic Physiol Opt*; **28**(6):503-23.
- Bo S., Pisu E. (2008). Role of dietary magnesium in cardiovascular disease prevention, insulin sensitivity and diabetes. *Curr Opin Lipidol*; **19**(1):50-6.
- Champagne C.M. (2008). Magnesium in hypertension, cardiovascular disease, metabolic syndrome, and other conditions: a review. *Nutr Clin Pract*; **23**(2):142-51.
- Chiuve S.E., Korngold E.C., Januzzi J.L., Gantzer M.L., Albert C.M. (2011). Plasma and dietary magnesium and risk of sudden cardiac death in women. *Am J Clin Nutr*; **93**(2):253-60.
- Ciarallo L., Brousseau D., Reinert S. (2000). Higher-dose intravenous magnesium therapy for children with moderate to severe acute asthma. *Arch Ped Adol Med*; **154**(10):979-83.
- Fallico R., Sciacca S., Bellassai S., Ferrante M. (1984). Risultati di un'indagine sulla concentrazione di fluoro nelle acque potabili della provincia di Catania. *Giornale di Igiene e Medicina Preventiva*. **25**(1):110-15.
- Housecroft C.E., Sharpe A.G. (2008). Inorganic Chemistry (3rd ed.). Prentice Hall. pp. 305–6.
- Kuo H.W., Chen. P.S., Ho. S.C., Wang L.Y., Yang C.Y. (2010). Trihalomethanes in drinking water and the risk of death from rectal cancer: does hardness in drinking water matter? *J Toxicol Environ Health A*. **73**(12):807-18.
- Turekian K.K. (1968).Oceans. Prentice-Hall.
- Wester PO. (1987). Magnesium. *Am J Clin Nutr*, **45**:1305-12.
- WHO (World Health Organization). (2011). Guidelines for Drinking-water Quality. 4th ed. WHO, Geneva, Switzerland. ISBN 978 92 4 154815 1.

Chapter 24
Manganese (Mn)

Margherita Ferrante.

24.1 ENVIRONMENTAL EFFECT

Manganese, atomic number 25, occurs as oxides and hydroxides in soils, and it is one of the most abundant metals in soil. Manganese is an essential element for all species. Some organisms, such as diatoms, molluscs and sponges, accumulate high amount of manganese. Fish can have up to 5 ppm and mammals up to 3 ppm in their tissue, although normally they have around 1 ppm (U.S.EPA, 2004).

Manganese is essential to iron and steel production. Manganese is a key component of low-cost stainless steel. Manganese dioxide is also used as a catalyst. Manganese is used to decolorize glass and make violet coloured glass. Potassium permanganate is a potent oxidizer and used as a disinfectant. Other Mn compound that find application are manganese oxide in fertilizers and ceramics (U.S.EPA, 2004).

24.2 EFFECT ON HUMAN HEALTH

Manganese is an essential nutrient needed for proper functioning of human body. Manganese is a constituent of several metalloenzymes and acts as an enzyme co-factor (i.e. superoxide dismutase, glutamine synthetase, arginase, but also oxidoreductases, transferases, hydrolases, lyases, isomerases, and ligases). It is an important compound for enzymes implicated in carbohydrate, amino acids, proteins, lipid and sterol metabolism and oxidative phosphorylation generating cellular energy-ATP (Fraga, 2005; WHOa, 2011).

Manganese is involved in the function of numerous organ systems and is needed for normal immune function, regulation of blood sugars, production of cellular energy, reproduction, digestion, and bone growth. Manganese works with vitamin K to support blood clotting. As a component of superoxide dismutase (SOD), Mn has important antioxidant properties since manganese-SOD (Mn-SOD) is one of the body's main front-line defense mechanisms against damaging free radicals (WHOa, 2011).

The main source of manganese in the general population is food. Foods like nuts, grains, fruits, legumes, tea, leafy vegetables, infant formulas, and some meat and fish are rich in manganese (ATSDR, 2000; U.S. EPA, 2004). Daily consumption of manganese originating from food range between 0.7 to 10.9 mg/day and can be even higher among vegetarians and heavy tea drinkers (ATSDR 2010; Greger 1999).

Manganese has been associated with the toxic effects listed in Table 24.1.

Table 24.1 Effect of Manganese (WHOa, 2011)	
Cardivascular effects	Heart damage in rat
Reproductive effects	Increased infant mortality, degeneration of testes
Neurological effects	Neurological deficits, damage of pathway of neurotrasmettitors, hallucinations, Parkinson-like symtomps, hyperactivity in children, increased of violent acts.
Kidney and Liver effects	Nephrithis and cirrhosis in rats

Manganese is also an essential mineral needed for proper foetal development. Manganese deficiency in humans appears to be rare, because manganese is present in many common foods. The findings of two recent studies (Vigeh et al., 2008; Zota et al., 2009) indicate that lower maternal blood manganese is associated with foetal intrauterine growth retardation (IUGR) and lower birth weight (Wood, 2009). A deficiency in manganese intake, also, causes seizure activity, leads to poor bone formation, impairs fertility, and causes birth defects in humans; it can also manifest as transient dermatitis, hypocholesterolemia, and an increased liver enzyme (alkaline phosphatase) levels (U.S.EPA, 2004; WHOa, 2011).
There are differences in manganese intake between breast-fed and formula-fed infants up to 6 month of age due to different concentrations of manganese in human milk and infant formula. Human milk contains on average 3.5 to 7.5 µg/l manganese and infant formula 100 times higher concentrations of this metal (Ljung & Vahter, 2007). Consequently, the similar ratio exists in estimated daily weight-adjusted manganese intake in infants assuming an average weight of 6 kg for an infant aged 6 months and an intake of 780 ml of milk per day (IOM 2001). After the introduction of solid foods, the contribution of manganese intake from milk decreases (U.S. EPA 1996a). Manganese concentration in the drinking water of the mixing formula should be taken into consideration when the infant's daily manganese exposure is assessed.
According to Nutrition Board of the Institute of Medicine an adequate intake (AI) value for manganese is 3 µg/l for infants 0-6 months old, which is lower end of ingested Mn/day from human milk. For adult men and women AI is 2.3 mg Mn/day and 1.8 mg Mn/day, respectively. The manganese toxic effects depend on the exposure level, route of exposure, chemical form, the age, nutritional, health status and individual variation.
The uptake is regulated so that when dietary manganese levels are high, the gastrointestinal absorption is reduced (Ljung & Vahter 2007). The main route of manganese excretion is through the biliary system. Infants and young children are more susceptible to high manganese levels due to the incompletely developed biliary system, greater retention and higher affinity of certain tissue sites for manganese than adults (Lonnerdal, 1994). A small part of absorbed manganese is excreted through urine, sweat, hair and human milk (U.S.EPA, 2004).
Manganese effects occur mainly in the respiratory tract and in the brains. Symptoms of manganese poisoning are hallucinations, forgetfulness and nerve damage. Manganese can also cause Parkinsonism, lung embolism, bronchitis. When men are exposed to manganese for longer periods of time they may become impotent. A syndrome that is caused by manganese has symptoms such as schizophrenia, dullness, weak muscles, headaches and insomnia. Because manganese is an essential element for human health shortages of manganese can also cause health effects such as: fatness, glucose intolerance, blood

clotting, skin problems, lowered cholesterol levels, bone disorders, birth defects, changes of hair colour and neurological symptoms.

Occupational studies on very high levels of inhaled manganese compounds have shown sufficient evidence of the occurrence of the syndrome known as "manganism", that is a set of neurological signs and symptoms which are very similar to Parkinson disease (PD). It is characterized by weakness, anorexia, muscle pain, apathy, slow speech, monotonous tone of voice, emotionless "mask-like" facial expression, and slow clumsy movements of the limbs. Manganese also affects lung, liver and cardiovascular system and induces reproductive and foetal toxicity (USEPA, 2004).

Children are more vulnerable to manganese adverse health effects because of their sensitive nervous system, their low bile excretion, high gastrointestinal absorption and homogenous diet. Adverse effects are manifested primary as behavioural disorders and intellectual impairment. In Bangladesh a study on drinking water found an association between manganese exposure and neurotoxic effects with an average of 800 µg Mn/l and a dose-response association between water manganese concentrations and test scores of performance as well as verbal ability (Wasserman et al. 2006). Woolf et al. (2002) reported that a 10-year old boy with abnormal verbal and visual memory function had elevated serum (0.90 µg/dL vs. normal value of < 0.265 µg/dL), whole blood, urine, and hair manganese concentrations following chronic ingestion of well water containing moderately elevated levels (~1.2 ppm) of manganese (Manganese Health Research Program, USA). Differences in behaviour (hyperactivity, oppositional behaviour) were shown in a study including two groups of children with average 600 µg/l and 160 µg/l manganese in the tap water, respectively (Bouchard et al. 2007). Recently, a cross-sectional study involving 362 children, 6–13 years of age in Southern Quebec, Canada, reported a 10-fold increase in water manganese to be associated with a decrease of 2.4 IQ points at manganese levels common in groundwater drinking water sources (median Mn level 34 µg/l) and below current drinking water standards of 50 µg/l (Bouchard et al., 2011). Nevertheless, more studies will be needed to confirm these findings (Chen & Copes 2011). Experimental animal data, especially rodent data, are not appropriate for human risk assessment because the physiological requirements for manganese vary among different species (WHOa, 2011).

The mechanism for the neurodegenerative damage specific for selected brain regions is not clearly understood. Manganese neurotoxicity may be associated with its interaction with other essential trace elements, including iron, zinc, copper and aluminium. The possible mechanism of manganese toxicity can be disturbance in iron homeostasis and the valence state of manganese. Studies in cultures cells have shown increase in cellular iron uptake when cultured cells were exposed to manganese compounds. The excess accumulation of iron in neurons may consequently produce the cellular oxidative stress that leads to neuronal damage (Crossgrove & Zheng, 2004).

On the other hand, recent evidence from human and non-human primate studies provides strong support to the hypothesis that excess levels of manganese accumulation in the brain results in an atypical form of Parkinsonism that is not associated with the degeneration of nigrostriatal dopaminergic neurons as it is the case in Parkinson disease. This new findings suggest that the Mn-induced movement abnormalities are not consequences of the decreased synthesis or concentration of dopamine, but rather a problem of the ability to release the available dopamine. Further investigations are needed to confirm this new suggested mechanism of manganese neurotoxicity (Guilarte, 2010).

In the EC directive on the quality of water intended for human consumption (98/83/EC) the value for manganese is set at 50 µg/l and represents an indicative value not legally binding (EU, 1998).
The Lifetime Health Advisories represents that portion of an individual's total exposure that is attributed to drinking water and is considered protective of noncarcinogenic adverse health effects over a lifetime exposure. The reference dose (RfD) of 0.14 mg Mn/kg-day (U.S.EPA, 1997) is estimated to be an intake for the general population that is not associated with adverse health effects; this is not meant to imply that intakes above the reference dose are necessarily associated with toxicity (EPA, 2004).
The World Health Organization health-based guideline value for drinking water was 400 µg Mn/l and the tolerable daily intake (TDI) was 60 µg Mn/kg body weight (WHO, 2004b; WHO, 2006). In a recent edition, WHO published that "this health-based value is well above concentrations of manganese normally found in drinking-water; it is not considered necessary to derive a formal guideline value" (WHOb, 2011).

References:

- ATSDR (2010). Addendum to the Toxicological Profile for Manganese. Agency for Toxic Substances and Disease Registry, Division of Toxicology and Environmental Medicine, Atlanta.
- Bouchard M., Laforest F., Vandelac L., Bellinger D., Mergle D. (2007). Hair Manganese and Hyperactive Behaviors: Pilot Study of School-Age Children Exposed through Tap Water. *Environ Health Perspect*, **115**:122–27.
- Bouchard M.F., Sauvé S., Barbeau B., Legrand M., Brodeur M.È., Bouffard T., Limoges E., Bellinger D.C., Mergler D. (2011). Intellectual Impairment in School-Age Children Exposed to Manganese from Drinking Water. *Environ Health Perspect*, **119**(1):138–43.
- Chen H., Copes R. (2011). Manganese in Drinking Water and Intellectual Impairment in School-Age Children. *Environ Health Perspect*. **119**(6): A240–41.
- Crossgrove J., Zheng W. (2004). Manganese toxicity upon overexposure. *NMR in Biomedicine*, **17**(8):544–53.
- Fraga C.G. (2005). Relevance, essentiality and toxicity of trace elements in human health. *Molecular Aspects of Medicine*, **26**:235–44.
- Greger J.L. (1999). Nutrition versus toxicology of manganese in humans: evaluation of potential biomarkers. *Neurotoxicology*, **20**:205–12.
- Guilarte R.T.(2010). Manganese and Parkinson's Disease: A Critical Review and New Findings. *Environ Health Perspect*, **118**(8):1071–80.
- Ljung K., Vahter M. (2007). Time to Re-evaluate the Guideline Value for Manganese in Drinking Water? *Environ Health Perspect*; **115**(11):1533–38.
- Lönnerdal B. (1994). Manganese nutrition of infants. In: Klimis-Tavantzis D, editor. Manganese in health and disease. Boca Raton (FL): CRC Press; 2004. p. 175–91.
- Mitchell E., Frisbie S., Sarkar B. (2011). Exposure to multiple metals from groundwater a global crisis: Geology, climate change, health effects, testing, and mitigation. *Metallomics*, **3**:874–908.

- U.S.EPA. (2004). Drinking Water Health Advisory for Manganese. U.S. Environmental Protection Agency. EPA-822-R-04-003 January, 2004.
- Vigeh M., Yokoyama K., Ramezanzadeh F., Dahaghin M., Fakhriazad E., Seyedaghamiri Z., Araki S. (2008). Blood manganese concentrations and intrauterine growth restriction. *Reprod Toxicol*, **25**(2):219-23.
- Wasserman G.A., Liu X., Parvez F., Ahsan H., Levy D., Factor-Litvak P., Kline J., Van Geen A., Slavkovich V., Lolacono N.J., Cheng Z., Zheng Y., Graziano J.H. (2006). Water manganese exposure and children's intellectual function in Araihazar, Bangladesh. *Environ Health Perspect*, **114**:124–29.
- WHO. (2004). Guidelines for Drinking-water Quality. Recommendations. Third Edition. Volume 1, World Health Organization. Genève.
- WHOa. (2011). Manganese in Drinking-water. Background document for development of WHO Guidelines for Drinking-water Quality. World Health Organization, Genève.
- WHOb (World Health Organization). (2011). Guidelines for Drinking-water Quality. 4th ed. WHO, Geneva, Switzerland. ISBN 978 92 4 154815 1.
- Wood RJ. (2009). Manganese and birth outcome. *Nutr Re*, **67**(7):416-20.
- Zota A.R., Ettinger A.S., Bouchard M., Amarasiriwardena C.J., Schwartz J., Hu H., Wright R.O. (2009). Maternal blood manganese levels and infant birth weight. *Epidemiology*, **20**(3):367-73.
- 98/83/EC, 1998. Council Directive (98/83/EC) of 3 November 1998, relating to the quality of water intended for human consumption. Off. J. Eur. Commun.

Chapter 25
Mercury (Hg)

Cristina Mauceri, Zorica Rasic-Milutinovic, Dragana Jovanovic.

25.1 ENVIRONMENTAL EFFECT

Mercury, atomic number 80 is a naturally occurring metal, which has several forms. Mercury is the only metal, which is liquid at ordinary temperatures and pressure and appears as a dense shiny, silver-white odorless liquid. If heated, it forms a colorless, odorless gas. Mercury combines with other elements, such as chlorine, sulfur, or oxygen, to form inorganic mercury compounds or "salts".

Release of mercury to air, water and through products may cause adverse impacts on environment and may pose significant risks to aquatic life and ultimately to human beings due to contamination of food chain. Metallic mercury may also get converted to inorganic and organic mercury, in fact, mercury also combines with carbon to make organic mercury compounds, the most common of which, methylmercury, is produced mainly by microbiological organisms in the water and soil. Mercury is present in the inorganic form in surface water and groundwater at concentrations usually below 0.5 µg/l, about a thousand times lower than "safe" drinking water standards, although local mineral deposits may produce higher levels in groundwater (WHO, 2011).

Metallic mercury is used to produce chlorine gas and caustic soda, and is also used in thermometers, dental fillings, and batteries. Mercury salts are sometimes used in skin lightening creams and as antiseptic creams and ointments. Food is the main source of Hg in general population through consumption of fish and shellfish. The average daily intake of Hg from food is in the range 2–20 µg (WHO, 2005). People can also be exposed to mercury by breathing vapors in air from spills, incinerators, and industries that burn mercury-containing fuels.Exposure to contaminated workplace air or skin contact in the workplace (dental, health services, chemical, and other industries) and through the release of mercury from dental work and medical treatments can contribute to Mercury load in the body (ATSDR, 1999). Mercury enters the environment as a result of normal breakdown of minerals in rocks and soil through exposure to wind and water. Release of mercury from natural sources has remained fairly constant over the years. However, mercury concentrations in the environment are increasing; this is ascribed to human activity. Most of the mercury released from human activities is released into air, through fossil fuel combustion, mining, smelting and solid waste combustion. Some forms of human activity release mercury directly into soil or water, for instance the application of agricultural fertilizers and industrial wastewater disposal. All mercury that is released in the environment will eventually end up in soils or surface waters.

The guideline value of WHO for mercury in drinking-water is 0.006 mg/l (WHO, 2011).

25.2 EFFECT ON HUMAN HEALTH

Mercury has many effects on human health (see Table 25.1), depending on its chemical form, the dose, the age of the person exposed (the fetus is the most susceptible), the duration of exposure, the route of exposure - inhalation, ingestion, dermal contact and the health of the person exposed. Absorption is about 80% for mercury vapor and nearly 100% for oral absorption. It is primarily distributed in the kidneys and brain and readily transferred to the fetus via the placenta. For this reason mercury could be responsible for birth defects and miscarriages. It is eliminated via the urine, feces, expired air, and breast milk.

A number of studies have shown that mercury binds to DNA and can cause strand breaks *in vitro* (IPCS, 2003), with weak genotoxic activity but without causing point mutations (WHO, 2005) and with a negative effect on sperm cells.

The nervous system is very sensitive to all forms of mercury (see Table 25.1). Methylmercury and metallic mercury vapors are more harmful than other forms, because more mercury in these forms reaches the brain. Mercury has a potential to induce neurotoxicity years after cessation of exposure or as a result of low-level exposure over a large portion of the life span. The health effects from mercury, in particular methylmercury, upon infants and children depend on the dose, with severe symptoms presenting with exposure to doses of 100 μg/kg/day, mild symptoms with greater than 10 μg/kg/day, and sub-clinical symptoms with less than 1 μg/kg/day (Hyman M, 2004). Symptoms include late development in walking and talking, decreased performance on neurological tests, appearance of neurodevelopmental disorders, intellectual retardation and autism (Weiss, 2000). Mercury neurotoxicity in adult subjects includes encephalopathy, peripheral neuropathy, Parkinsonian symptoms, tremor and ataxia, impaired hearing, tunnel vision, dysarthria, headache, fatigue, impaired sexual function, and depression (Kakkar & Jaffery, 2005). Other Central Nervous System (CNS) toxicity includes erethism with symptoms of shyness, emotional lability, nervousness, insomnia, memory impairment, and inability to concentrate. Acute exposure to high concentration of mercury can cause lung symptoms ranging from irritation to respiratory failure and damage kidneys, skin and gastrointestinal tract. Renal toxicity includes proteinuria, renal syndrome, and acute renal failure. Gastrointestinal symptoms include nausea, vomiting, diarrhoea, and colitis. Dermal toxicity includes allergic dermatitis, chelitis, gingivitis, stomatitis, and excessive salivation. No human data indicate that exposure to any form of mercury causes cancer, but the human data currently available are very limited (EPA, 2005).

Table 25.1 Mercury toxic effects (WHO, 2005)	
Organs	**Effects**
Reproductive system	Impaired sexual function, teratogenic effect in reproduction.
Kidney	Proteinuria, renal syndrome, acute renal failure.
Respiratory system	Irritation, respiratory failure
Nervous system	Deafness, muscle in-coordination, loss of sensation and difficulties with memories, dyslexia, attention deficit, hyperactivity disorder, encephalopathy, peripheral neuropathy, parkinsonian symptoms, tremor and ataxia, tunnel vision, dysarthria, headache, fatigue, depression, emotional lability, nervousness, insomnia, memory impairment, inability to concentrate.
Skin	Allergic dermatitis, chelitis, gingivitis, stomatitis.
Visual system	Changes in vision.

References:

- ATSDR. (1999). – ToxFAQs. Mercury. Agency for Toxic Substances and Disease Registry.
- EPA – U.S. Environmental Protection Agency (2005). Cancer guidelines.
- Hyman M. (2004). The Impact of Mercury on Human Health and the Environment. *Alternative therapies*; **6**(10):70-5.
- IPCS. (2003). Concise International Chemical Assessment Document 50: Elemental mercury and inorganic mercury compounds: human health aspects. Geneva, World Health Organization and International Programme on Chemical Safety.
- Kakkar P., Jaffery F.N. (2005). Biological markers for metal toxicity. *Environmental Toxicology and Pharmacology*; **19**:335–49.
- Weiss B. (2000). Vulnerability of children and the developing brain to neurotoxic hazards. *Environ. Health Perspect*; **108**:375–81.
- WHO (World Health Organization). (2011). Guidelines for Drinking-water Quality. 4th ed. WHO, Geneva, Switzerland. ISBN 978 92 4 154815 1.
- WHO (World Health Organization). (2005). Mercury in drinking-water. Background document for development of WHO Guidelines for Drinking-water Quality. Geneva.

Chapter 26
Nickel (Ni)

Gea Oliveri Conti, Zorica Rasic-Milutinovic, Dragana Jovanovic.

26.1 ENVIRONMENTAL EFFECT

Nickel atomic number 28, is a hard, malleable, and ductile metal. It is a good conductor of heat and electricity. The major use of nickel is in the preparation of alloys while the remainder of consumption is divided between steels, rechargeable batteries, catalysts and other chemicals, coinage, foundry products, and plating. Humans use nickel for many different applications such as alloys and products such as jewellery.

Nickel is resistant to corrosion also by sea water indeed it is used in the construction of metal elements in boats and desalination plants. Nickel is a compound that occurs in the environment only at trace levels. Most nickel on earth is inaccessible because it is locked away in the planet's iron-nickel molten core (10%). Nickel is found in all soil (from 0.2 ppm to 450 ppm, the average is any case about 20 ppm; ATSDR, 2005) and is emitted from volcanoes; it is dissolved in the sea bottom (8 billion tonnes).

Foodstuffs naturally contain small amounts of nickel except chocolate and fats. Plants accumulate nickel, as a result nickel uptake from food will be eminent; also tobacco smoking exposes to high absorption of nickel, therefore, people may be exposed to nickel through air, drinking water, food or tobacco. Water generally contributes 0.005–0.025 mg daily (i.e., 2–11% of the total daily oral intake of nickel) (WHO, 2005).

Typical daily intake from drinking water is 2 µg, however absorption of soluble nickel compounds from drinking-water is higher than from food (WHO, 2011).

The guideline value of WHO for nickel in drinking-water is 0.07 mg/l (WHO, 2011).

Reference values for nickel in healthy adults are 0.2 µg/L in serum and 1–3 µg/L in urine (ASTDR, 2005).

The EPA recommends that drinking water should contain no more than 0.1 mg/L of nickel, while the Italian law is more protective, setting a limit of 20 µg/L.

Skin contact with nickel-contaminated soil or water may also result in nickel exposure. Nickel is essential in trace quantities but when the uptake is too high it can cause disease to humans.

The primary source of nickel in drinking-water is leaching from metal pipes and fittings in contact with drinking-water, however, nickel may also be present in groundwaters as a consequence of dissolution of nickel from rocks. Nickel concentrations in groundwater depend on pH and soil, moreover the acid rain increases the mobility of nickel from soil to groundwater.

Higher concentrations in drinking water were found after water was left overnight in plumbing fittings plated with chromium on a base of nickel, in water boiled in electric kettles depending on the material of the heating element and when the pipes were assembled with tinned copper and gunmetal fittings (WHO, 2005). However, these leachings decrease significantly with time (EU, 2004).

26.2 EFFECT ON HUMAN HEALTH

Excessive uptake of nickel has the following consequences:

- Higher risk of development of lung, nose, larynx and prostate cancers .
- Neurological effects after exposure to nickel gas.
- Lung embolism.
- Respiratory failure.
- Teratogenic effects.
- Asthma and chronic bronchitis.
- Allergic reactions from jewellery (skin rashes).
- Heart disorders.

The primary targets of toxicity after nickel ingestion appear to be the immune, and the reproductive systems, and developing organisms. The most commonly reported adverse health effect associated with nickel exposure is contact dermatitis. After an individual becomes sensitized to nickel, dermal contact with a small amount of nickel or oral exposure to fairly low doses of nickel can result in dermatitis. Approximately 10–20% of the general population is sensitized to nickel. Inconsistent results of animal studies have been reported for the reproductive toxicity (decreased sperm motility and count, sperm abnormalities, decreased fertility) and developmental effects (fetal loss and decreased survival) of nickel. Human data are lacking (Mitchell et al., 2011; ASTDR, 2005).

At the cellular level, it increases oxidative products and proteins (Mitchell et al., 2011).

A number of studies have examined the genotoxicity of various forms of nickel and nickel compounds. Mixed results for frequency of gene mutations in different Chinese hamster cells systems have been found. Thus, an increase in gene mutation frequency has also been found in Chinese hamster ovary AS52 cells (*grp* locus), mouse lymphoma cells and virus-infected mouse sarcoma cells (ASTDR, 2005). There is some evidence that nickel caused genetic damage, including DNA strand breaks, mutations, chromosomal damage, cell transformation, and disrupted DNA repair. In WHO document (2005) comprising nickel in drinking water, TERA (1999) conclusion, as part of the EU (2004) nickel sulphate risk assessment, was cited as "*evidence for genotoxicity is mixed, although water soluble nickel compounds have been generally consistent in inducing effects in certain kinds of mammalian assays, particularly mutagenic responses and DNA damage in vitro, chromosomal effects including aberrations and sister-chromatid exchanges in vitro and in vivo, and carcinogenic transformation of mammalian cells in vitro. Responses in many of these assays were weak and occurred at toxic doses.*"

The International Agency for Research on Cancer (IARC) has listed nickel compounds within group 1 and nickel within group 2B. OSHA does not regulate nickel as a carcinogen. Nickel is on the ACGIH Notice of Intended Changes as a Category A1, confirmed human carcinogen.

Workers who drank water containing high amounts of nickel showed stomach ache and adverse effects to their blood and kidneys.

Tumors of the lung and nasal sinus have been found in refinery workers. The EPA has determined that nickel refinery dust and nickel subsulfide are human carcinogens.

Exposures of the population from sources such as foods and drinking water, are almost always too low to be of concern.
The larger part of all nickel compounds that are released in the environment will be adsorbed in sediment or soil particles but in acidic ground nickel is more mobile and it will often rinse out to the groundwater.
The larger part of all nickel compounds that are released to the environment will adsorb to sediment or soil particles but in acidic ground nickel is bound to become more mobile and it will often rinse out to the groundwater.

References:

- Agency for Toxic Substances and Disease Registry (ATSDR). (2005). *Toxicological Profile for Nickel* (Update). Atlanta, GA: U.S. Department of Public Health and Human Services, Public Health Service.
- Decreto Legislativo 2 febbraio 2001, n. 31, "Attuazione della direttiva 98/83/CE relativa alla qualità delle acque destinate al consumo umano". Italy.
- Mitchell E., Frisbie S., Sarkar B. (2011). Exposure to multiple metals from groundwater a global crisis: Geology, climate change, health effects, testing, and mitigation. Metallomics, **3**:874–908.
- U.S. Cancer Statistics Working Group. United States Cancer Statistics:1999–2007. (2010). Incidence and Mortality Web-based Report. Atlanta (GA): Department of Health and Human Services, Centers for Disease Control and Prevention, and National Cancer Institute. Available at: http://www.cdc.gov/uscs. (accessed on 30 July 2012).
- EU. Nickel sulphate risk assessment. Draft, May 2004. Prepared by the Danish Environmental Protection Agency for the European Union, 2004.
- WHO (World Health Organization). (2011). Guidelines for Drinking-water Quality. 4th ed. WHO, Geneva, Switzerland. ISBN 978 92 4 154815 1.
- World Health Organization (WHO). (2005). Nickel in Drinking-water. Background document for development of WHO - Guidelines for Drinking-water Quality. Geneve.
- Toxicology Excellence for Risk Assessment (*TERA*). (1999). Toxicological review of soluble nickel salts. Research Triangle Park, NC, Toxicology Excellence for Risk Assessment.

Chapter 27
Potassium (K)

Gea Oliveri Conti, Maria Fiore.

27.1 ENVIRONMENTAL EFFECT

Potassium, atomic number 19, is a alkali metal and the seventh most common element on earth (Lewis, 1997).

Potassium is an essential element in humans and is seldom, if ever, found in drinking water at levels that could be a concern for healthy humans. It occurs widely in the environment, including all natural waters. It can also occur in drinking-water as a consequence of the use of potassium permanganate as an oxidant in water treatment. In some countries, potassium chloride is being used in ion exchange for household water softening in place of, or mixed with, sodium chloride, so potassium ions would exchange with calcium and magnesium ions (WHO, 2009).

Potassium is present in all animal and plant tissues. The primary source of potassium for the general population is the diet, as potassium is found in all foods, particularly vegetables and fruits (WHO, 2009).

Potassium plays a critical role in many vital cell functions, such as metabolism, growth, repair and volume regulation, as well as in the electric properties of the cell (Adriogué & Wesson, 1994).

27.2 EFFECT ON HUMAN HEALTH

Health effects due to potassium consumption in drinking water are unlikely to occur in most individuals. Potassium intoxication by ingestion is rare, because potassium is rapidly excreted in the absence of pre-existing kidney damage and because large single doses usually induce vomiting (Gosselin et al., 1984). However, acute ingestion of doses greater than 2.0 meq/kg bw (>78 mg/kg bw or 5.5 g for a 70 kg adult) by individuals with normal kidney function can overwhelm homeostatic mechanisms and possibly cause death (Buckley et al., 1995).

Adverse effects may also occur when K^+ plasma concentrations are lower (hypokalemia) than the normal range (3.5–5.0 mmol/L). Both hyperkalemia and hypokalemia result from disruptions in transcellular homeostasis or in the renal regulation of K^+ excretion (Gennari, 2002).

As indicated above, potassium is an essential element for human nutrition. Potassium and sodium maintain the normal osmotic pressure in cells. Potassium is a cofactor for many enzymes and is required for the secretion of insulin, creatinine phosphorylation, carbohydrate metabolism and protein synthesis.

Excessive loss of salts, such as through severe diarrhoea or intense and prolonged sweating, can result in a loss of potassium, which can result in hypokalaemia if the loss is sufficient. This can cause a range of effects, including cardiac arrhythmia, muscle weakness, nausea and vomiting, and low muscle tone in the gut. Longer-term hypokalaemia is believed to cause a predisposition to hypertension (UKEVM, 2003).

Case-studies of toxicity resulting from high doses of salt substitutes have described chest tightness, nausea and vomiting, diarrhoea, hyperkalaemia, shortness of breath and heart failure (Restuccio, 1992). When hyperkalemia occurs, it causes a partial depolarization of the cell membrane by altering the ratio of intracellular to extracellular fluid K^+ concentration, which in turn affects neural transmission, muscle contraction and blood vessel constriction (IOM, 2005).

Although concentrations of potassium normally found in drinking-water are generally low and do not pose health concerns, the high solubility of potassium chloride and its use in treatment devices such as water softeners can lead to significantly increased exposure.

Although potassium may cause some health effects in susceptible individuals, potassium intake from drinking-water is well below the level at which adverse health effects may occur. Health concerns would be related to the consumption of drinking water treated by potassium-based water treatment affecting only individuals in the high risk groups (individuals with kidney dysfunction or other diseases such as heart disease, coronary artery disease, hypertension, diabetes, adrenal insufficiency, pre-existing hyperkalaemia; people taking medications that interfere with normal K-dependent functions in the body; and older individuals or infants) (Guidance, 2008).

References:

- Adriogué H.J., Wesson D.E. (1994). Blackwell's basics of medicine: Potassium. Blackwell Scientific Publications, Boston, MA.
- Buckley N.A., Dawson A.H., Reith D.A. (1995). Controlled release drugs in overdose: clinical considerations. *Drug Saf.*, **12**(1):73–84.
- Gennari J.F. (2002). Disorders of potassium homeostasis: hypokalemia and hyperkalemia. *Crit. Care Clin.*, **18**(2):273–88.
- Gosselin R.E., Smith R.P., Hodge H.C. (1984). Clinical toxicology of commercial products. 5th edition. Williams & Wilkins, Baltimore, MD.
- Guidance on Potassium from Water Softeners Federal-Provincial-Territorial Committee on Drinking Water, Federal-Provincial-Territorial Committee on Health and the Environment, Ottawa, Ontario May 2008 [Health Canada (2008). Guidance on Potassium from Water Softeners. Water, Air and Climate Change Bureau, Healthy Environments and Consumer Safety Branch, Health Canada, Ottawa, Ontario.
- Restuccio A. (1992). Fatal hyperkalaemia from a salt substitute. *American Journal of Emergency Medicine*, **10**:171–73.
- UKEVM. (2003). Risk assessments: Potassium. In: Safe upper levels for vitamins and minerals. London, United Kingdom Food Standards Agency, Expert Group on Vitamins and Minerals, p.299. (http://cot.food.gov.uk/pdfs/vitmin2003.pdf) (accessed on 5 July 2012).
- WHO. (World Health Organization) (2009). Potassium in drinking-water. Guidelines for Drinking-water Quality. Geneva.

Chapter 28
Radium (Ra)

Maria Fiore, Caterina Ledda.

28.1 ENVIRONMENTAL EFFECT

Radium (Ra) is a naturally-occurring silvery white radioactive metal that can exist in several forms called isotopes. The main isotopes are ^{224}Ra (half-life 3.6 days), ^{226}Ra (half-life 1620 years) and ^{228}Ra (half-life 5.8 years).
It is formed when uranium and thorium (two natural radioactive substances) decay (break down) in the environment. Radium has been found at very low levels in soil, water, rocks, coal, plants and food. You may be exposed to higher levels of r adium if you live in an area where it is released released into the air from coal or other fuels burning, or if your drinking water is taken from a source that contains a big amount of natural radium, such as a deep well, or from a source near a radioactive waste disposal site. Levels of radium in public drinking water are usually less than one picocurie per liter (about one quart), although higher levels (more than 5 picocuries per liter) have been found (ATSDR, 1990). The most significant water-related releases of radium may be from the leaching of uranium mine tailings and from the release of ore -processing effluents generated by leaching, decantation, and filtration processes. Radium in water may be readily adsorbed by sediments, soils, and aquifer components. It has been experimentally demonstrated that radium can be absorbed by soils and sediments, ferric hydroxide and quartz, kaolinite and montmorillonite, and muscovite and albite (ATSDR, 1990).

28.2 EFFECT ON HUMAN HEALTH

In communities where wells are used, drinking water can be an important source of ingested radium. Radium is absorbed from food and water in the gastrointestinal tract. Although most is excreted, bones become the principal repository for retained radium because of the chemical similarity between radium and calcium. Radium can cross the placenta, and the ratio radium over calcium in the newborn reflects the ratio in the mother's blood. Limited evidence shows that under conditions of long-term intake the concentration of radium in the body is nearly invariant throughout life (Wrenn, 1985; Finkelstein, 1994).
Both ^{226}Ra and ^{228}Ra rapidly induce changes in bone structure and haematopoiesis. Bone sarcomas were found in all species tested within lifespan follow-up (Guse et al, 2002; ATSDR, 1990). Leukemias were reported following ^{224}Ra injection (ATSDR, 1990; Canu et al., 2011).

References:

- ATSDR. (1990). Toxicological profile for radium. Public Health Service, Agency for Toxic Substances and Disease Registry, U.S. Department of Health and Human Service.
- Canu I.G., Laurent O., Pires N., Laurier D., Dublineau I. (2011). Health Effects of Naturally Radioactive Water Ingestion: The Need for Enhanced Studies. *Environ Health Perspect*; 119:1676–80.
- Finkelstein M.M. (1994). Radium in drinking water and the risk of death from bone cancer among Ontario youths. *CMAJ*, **151**(5):565-71.
- Guse C.E., Marbella A.M., George V., Layde P.M. (2002). Radium in Wisconsin drinking water: an analysis of osteosarcoma risk. *Arch Environ Health*. **57**(4):294-303.
- Wrenn M.E., Durbin P.W., Howard B., Lipsztein J., Rundo J., Still E.T., Willis D.L. (1985). Metabolism of ingested U and Ra. *Health Phys*. **48**(5):601-33.

Chapter 29
Selenium (Se)

Marco Vinceti, Carlotta Malagoli, Margherita Ferrante.

29.1 ENVIRONMENTAL EFFECT

Selenium (Se), atomic number 34, is a metalloid found in all environmental matrices, including drinking water, soil, air and foods, and it is released from both natural sources (Se-containing rocks and soils and volcanic eruptions) and anthropogenic sources (coal combustion, industry of steel, ceramics, glass, electronics and petroleum, photography and pharmaceutical production, rubber vulcanizing, and irrigation of high-Se farmland soils). It can be naturally found at generally low concentrations in surface waters and ground waters all over the world, though some areas, not necessarily the seleniferous ones, may occasionally present unusually high levels of selenium (Hurtado-Jimenez & Gardea-Torresdey, 2007; Hudak, 2009; Kuisi & Abdel-Fattah, 2010).

Selenium may also be used in human medicine: as a pharmacological tool in anticancer therapy, as selenium sulfide for treatment of some dermatological diseases such as dandruff and seborrheic dermatitis of the scalp, and as a gamma-emitting isotope (^{75}Se) for radiological diagnostic purposes.

Selenium can exist as organic compounds (that are generally of interest for human nutrition) and as inorganic species, that are those generally found in drinking waters. The latter forms may be elemental selenium, selenite, selenate, selenious acid, selenic acid, selenium dioxide, selenium disulfide and other compounds.

29.2 EFFECT ON HUMAN HEALTH

From a biomedical and public health perspective, interest in selenium relies on both its nutritional and toxicological role, though neither effect has been entirely elucidated, and have generated strong controversies in the scientific community (Office of Drinking Water, 1990; WHO, 1996; Barceloux, 1999; Vinceti et al., 2009; OEHHA, 2010; Dennert et al., 2011; WHO, 2011; Gilron, 2012).

In fact, selenium is a cofactor of some enzymes which are of key importance for human health, such as the selenium-dependent glutathione-peroxidases and thioredoxin reductase, which play an important role in the defense against oxidative stress. On the other hand, selenium is also a toxic element, with a broad pattern of adverse effects associated not only to its dose but also to its chemical form (Aggett, 2010; Douron, 2010; Lee and Jeong, 2012). Even if other trace elements are also known to be both toxic and essential for human health, selenium has generated interest as a potential antioxidant and anti-carcinogen, following the results of some epidemiologic and laboratory studies (WHO Environmental Health Criteria, 1987). However, convincing evidence from recent clinical and experimental studies demonstrate that such alleged anticancer capacity was incorrect (Lippman et al., 2009) and that selenium itself is also pro-oxidant element in addition to having antioxidant properties associated to selenium-containing enzymes (Vinceti et al., 2009; Miller and Hontela, 2011; Nogueira and Rocha, 2011; Lee and Jeong, 2012; Medeiros et al., 2012). Selenium sulfide is also recognized as a carcinogen (US-EPA, 2009), and concern has been expressed about the environmental effects of contamination with selenium in agricultural and natural ecosystems (Hladun et al., 2012; Li et al., 2012).

While a large number of studies have investigated the relationship between selenium status, cancer risk and cardiovascular health, with mixed and intriguing results (Stranges et al., 2010; Dennert et al., 2011), surprisingly few investigations have examined the health effects of (inorganic) selenium from drinking water in humans (Vinceti et al., 2009). Studies carried out in the US found little evidence of any associations with signs and symptoms of disease in communities exposed to high- selenium levels in drinking water (Tsongas and Ferguson, 1977; Valentine et al., 1987; Valentine, 1997), with the possible exception of a higher prevalence of diarrhoea and of neurological diseases in the most exposed communities. On the converse, investigations carried out in an Italian setting where tap water with selenium content around 8 µg/l had been distributed suggested the possibility of adverse effects of such low-dose overexposure to drinking water selenium,such as an increased risk of melanoma and lymphoid malignancies as well as of amyotrophic lateral sclerosis (Vinceti et al., 2000b; Vinceti et al., 2010).

Little evidence of teratogenic effects of drinking water selenium was yielded by these studies (Tsongas and Ferguson, 1977; Vinceti et al., 2000a).

The majority of European Union countries and Canada, Australia, Japan, Thailand and New Zealand have set the upper standard for selenium at 10 µg/l, except for Russia (1 µg/l), US-EPA (50 µg/l) and California Environmental Protection Agency (30 µg/l) (OEHHA, 2010), and a background WHO document suggested as safe upper limit 40 µg/l. However, considering the very high toxicity of inorganic selenium species compared with organic ones, and the results of the few epidemiologic studies carried out on this topic, the 10 µg/l standard for drinking water selenium should be maintained or even reduced to adequately protect long term consumers of drinking waters with these amount of selenium, as well as susceptible populations.

The guideline value of WHO for selenium in drinking-water is 0.04 mg/l (WHO, 2011).

References:

- Aggett P.J. (2010). Toxicity due to excess and deficiency. *J Toxicol Environ Health A*; 73:175-80.
- Barceloux D.G. (1999). Selenium. *J Toxicol Clin Toxicol*; 37:145-72.
- Dennert G., Zwahlen M., Brinkman M., Vinceti M., Zeegers M.P., Horneber M. (2011). Selenium for preventing cancer. *Cochrane Database Syst Rev*, 5:CD005195.
- Douron M. (2010). U-shaped dose-response curves: implications for risk characterization of essential elements and other chemicals. *J Toxicol Environ Health A*, 73:181-86.
- Gilron G. (2012). The selenium drinking water quality guideline in Canada: the case for a re-evaluation. *Integr Environ Assess Manag*; 8:194-96.
- Hladun K.R., Parker D.R., Tran K.D., Trumble J.T. (2012). Effects of selenium accumulation on phytotoxicity, herbivory, and pollination ecology in radish (Raphanus sativus L.). *Environ Pollut*; 172C:70-5.

- Hudak P.F. (2009). Elevated fluoride and selenium in west Texas groundwater. *Bull Environ Contam Toxicol*; 82:39-42.
- Hurtado-Jimenez R., Gardea-Torresdey J. (2007). Evaluation of the exposure to selenium in Los Altos de Jalisco, Mexico. *Salud Publica Mex*; 49:312-15.
- Kuisi M.A., Abdel-Fattah A. (2010). Groundwater vulnerability to selenium in semi-arid environments: Amman Zarqa Basin, Jordan. *Environ Geochem Health*; 32:107-28.
- Lee K.H., Jeong D. (2012). Bimodal actions of selenium essential for antioxidant and toxic pro-oxidant activities: The selenium paradox (Review). *Mol Med Report*; 5:299-304.
- Li S., Xiao T., Zheng B. (2012). Medical geology of arsenic, selenium and thallium in China. *Sci Total Environ*; 421-22: 31-40.
- Lippman S.M., Klein E.A., Goodman P.J., Lucia M.S., Thompson I.M., Ford L.G, Parnes H.L., Minasian L.M., Gaziano J.M., Hartline J.A., Parsons J.K., Bearden J.D. 3rd, Crawford E.D., Goodman G.E., Claudio J., Winquist E., Cook E.D., Karp D.D., Walther P., Lieber M.M., Kristal A.R., Darke A.K., Arnold K.B., Ganz P.A., Santella R.M., Albanes D., Taylor P.R., Probstfield J.L., Jagpal T.J., Crowley J.J., Meyskens F.L. Jr, Baker L.H., Coltman C.A. Jr. (2009). Effect of selenium and vitamin E on risk of prostate cancer and other cancers: the Selenium and Vitamin E Cancer Prevention Trial (SELECT). *JAMA*; **301**(1):39-51.
- Medeiros M.C., Mello A., Gemelli T., Teixeira C., de Almeida M., de Andrade R.B., Wannmacher C.M., Guerra R.B., Gomez R., Funchal C. (2012). Effect of chronic administration of the vinyl chalcogenide 3-methyl-1-phenyl-2-(phenylseleno)oct-2-en-1-one on oxidative stress in different brain areas of rats. *Neurochem Res*; **37**(5):928-34.
- Miller L.L., Hontela A. (2011). Species-specific sensitivity to selenium-induced impairment of cortisol secretion in adrenocortical cells of rainbow trout (Oncorhynchus mykiss) and brook trout (Salvelinus fontinalis). *Toxicol Appl Pharmacol;* **253**:137-44.
- Nogueira C.W., Rocha J.B. (2011). Toxicology and pharmacology of selenium: emphasis on synthetic organoselenium compounds. *Arch Toxicol*; **85**:1313-59.
- OEHHA. (2010). Public Health Goal for selenium in drinking water. Office of Environmental Health Hazard Assessment, California Environmental Protection Agency, Oakland.
- Office of Drinking Water USEPA. (1990). The drinking water criteria document on selenium. Cleveland, OH: ICAIR Life System.
- Stranges S., Navas-Acien A., Rayman M..P, Guallar E. (2010). Selenium status and cardiometabolic health: state of the evidence. *Nutr Metab Cardiovasc Dis*; **20**:754-60.
- Tsongas T.A., Ferguson S.W. (1977). Human health effects of selenium in a rural Colorado drinking water supply. In: Hemphill DD, editor. In: Proceedings of *"Trace substances in environmental health-XI. A symposium"*. University of Missouri, Columbia, pp. 30-5.
- US-EPA. (2009). Selenium sulfide Quickview (CASRN 7446-34-6) - Carcinogenicity Assessment for Lifetime Exposure (Weight-of-Evidence Characterization).

- Valentine J.L. (1997). Environmental occurrence of selenium in waters and related health significance. *Biomed Environ Sci*; **10**:292-99.
- Valentine J.L., Kang H.K., Schluchter M. (1987). Effects on human health of exposure to selenium in drinking water. In: Combs GF, Levander OA, Spallholz JE, Oldfield JE, editors. In: Proceedings of "Selenium in Biology and Medicine - Part B". Van Nostrand Reihold Co, New York, 1987, pp. 675-87.
- Vinceti M., Bonvicini F., Rothman K.J., Vescovi L., Wang F. (2010). The relation between amyotrophic lateral sclerosis and inorganic selenium in drinking water: a population-based case-control study. *Environ Health;* 9:77.
- Vinceti M., Cann C.I., Calzolari E., Vivoli R., Garavelli L., Bergomi M. (2000a). Reproductive outcomes in a population exposed long-term to inorganic selenium via drinking water. *Sci Total Environ*; **250**:1-7.
- Vinceti M., Maraldi T., Bergomi M., Malagoli C. (2009). Risk of chronic low-dose selenium overexposure in humans: insights from epidemiology and biochemistry. *Rev Environ Health*; **24**:231-48.
- Vinceti M., Nacci G., Rocchi E., Cassinadri T., Vivoli R., Marchesi C., Bergomi M. (2000b). Mortality in a population with long-term exposure to inorganic selenium via drinking water. *J Clin Epidemiol*; **53**(10):1062-68.
- World Health Organization (WHO). (1996). *Guidelines for drinking water quality. Health criteria and other supporting information.* 2nd ed. Vol 2. Geneva, Switzerland.
- WHO (World Health Organization). (2011). Guidelines for Drinking-water Quality. 4th ed. WHO, Geneva, Switzerland. ISBN 978 92 4 154815 1.
- World Health Organization (WHO). (1987). Environmental Health Criteria. Selenium. Geneva, Switzerland.

Chapter 30
Silicon (Si)

Gea Oliveri Conti, Maria Cunsolo, Roberto Furnari.

30.1 ENVIRONMENTAL EFFECT

Silicon, atomic number 14, is not a metal. It is the second most abundant element in the Earth crust at 28 wt % (Exley, 1998; Sjöberg, 1996), but it is rarely found in its elemental form. Dissolution of Silicon, from soil minerals in water results in the formation, by hydrolysis, of soluble silica species (Iler, 1979).

Silicon may be released during weathering processes. It is also released under water during volcanic activity. Water in interspaces of marine sediments contains more silicon than the sea surface. The presence oft currenst causes silicon to flow from sediments to seawater. Antarctic weathering also releases Silicon.

Silicon is removed from waters naturally, through plankton fixation, sediment settling, or reactions of dissolved silicon with clay minerals (reverse weathering). Silicon is applied in alloy steel and its derivated products, chemical and electron industries, where it is processed under high temperatures. Industrially significant Silicon compounds – silicones - are rubber- or resin-like compounds applied as lubricants under high temperatures or as a sealing kit for many civil and industrial applications.

Silicone oils are applied in cosmetics, and for textile impregnation. In microchips this element is a semi conductor, as it is in transistors and other electronic parts. Solar panels consist of n-semi conductors of silicon and arsenic and p-semi conductors of silicon and boron.

Sand is the primary substance for commercially produced Silicon. Minerals such as talc, mica, feldspar, nepheline, olivine, vermiculite, perlite and kaolinite also contain Silicon.

Humans are exposed to numerous sources of silica/silicon including dust, pharmaceuticals, cosmetics and medical implants and devices, but the major and most important source of exposure for the majority of the population is the diet.

30.2 EFFECT ON HUMAN HEALTH

Silicon in drinking water derives from the weathering of rocks and soil minerals, therefore its concentration in water depends upon the surrounding geology. For example in the UK, Si concentrations range between 0.2-2.5 mg/L in the north and west of Britain, while they are much higher (2.8-14 mg/L) in the south and east of Britain. The Si concentration of European mineral waters is within a similar range (4-16 mg/L). Higher Si levels in drinking water (30-40 mg/L) have been reported in wasters from natural sources in Malaysia and Fiji (Jugdaohsingh, 2007). Drinking water represent the major source of Si in the diet. The major silica species present in drinking water is orthosilicic acid, that provides the most available source of Silicon to man. It is readily absorbed and at least 50% of intake excreted (Jugdaohsingh, 2007; Jugdaohsingh, 2002; Reffitt, 1999; Bellia, 1996).

In a experimental study on rats, Silicon supplementation was found to be more effective when rats were fed a low calcium diet (Carlisle, 1982), and a subsequent study showed that low dietary calcium enhanced the Silicon uptake (Nielsen, 1991). These results suggest that, either calcium and Silicon compete for the same absorption pathway, or that calcium forms insoluble, luminal calcium silicate that reduces silica bioavailability.

Magnesium could similarly reduce the bioavailability of Silicon by forming insoluble silicates, since magnesium orthosilicate is considered the predominant form of Silicon in urine and possibly in plasma (Berlyne, 1986). It has been suggested that Silicon controls the metabolism of calcium and magnesium (Charnot, 1971). Silicon is present as neutral orthosilicic acid that readily diffuses into erythrocytes and other tissues (Adler, 1986).

Tissue levels of Silicon vary. In the rat the highest levels are found in bone and other connective tissues such as, skin, nail, hair, trachea, tendons and aorta and very much less (10-20 fold less) in soft tissues (Exley, 1998). A similar tissue distribution of Silicon might be expected in humans, although this has not been investigated.

Silicon is thought to be integrally bound to connective tissues and their components, and to have an important structural role (Schwarz, 1973; Carlisle, 1972). Silicon deprivation studies have found detrimental effects on these tissues (Schwarz, 1972); the same effects are also thought to occur with normal ageing due to decline in Silicon levels in the tissues. Vice versa, Silicon supplementation has been reported to have beneficial effects on these tissues, especially on bones,. (Jugdaohsingh, 2004; Calomme, 2006; Barel, 2005; Spector, 2005).

Silicon is widely regarded as a non essential element in mammals, though it has been reported to have some beneficial actions, sometimes at rather high amounts.

Oral ingestion of crystalline or amorphous Silicon/silicates in the diet may also cause toxicity. Finely ground silicate minerals from eroded acid granite in drinking water has been linked to 'Endemic or Balkan Nephropathy', that is inflammation of the kidneys (interstitial nephritis), found in confined parts of the Balkans (Yugoslavia, Bulgaria and Romania) (Dobbie & Smith, 1984).

The role of Silicon in drinking water has not been much studied, and clear results have not yet emerged. Data published in 2005 (Gillette-Guyonnet et al., 2005) on the potential benefit of Silicon need to be confirmed and additional investigations to exclude causes of error related to methodological biases. If such an association does indeed exist, measures could be taken to reduce the incidence of Alzheimer disease (AD). Finally, it is possible that the causal role of aluminium in AD may have to be reconsidered due to the last paper published on this topic (Kawahara, 2005; Andrasi et al., 2005). It was especially shown that aluminium is implicated in the formation of neurofibrillary tangles (Walton, 2006). Moreover, it is possible that the neurotoxic effects of aluminium could be detected at a preclinical stage of AD giving us the opportunity to develop early treatment to prevent or retard the onset of AD or AD-like pathologies (Smorgon et al., 2004). In conclusion, this paper suggests that high Silicon concentrations in drinking water may protect against impairment of cognitive function. However, further studies are necessary to not only confirm these results but to also clarify the potential effect of Silicon against aluminum-induced neurotoxicity and the causal role of aluminum in AD.

During the 1990s, some investigations showed that Silicon (in its biologically available form, silicic acid or Si(OH)4), reduced the gastrointestinal absorption of dietary Al58 and facilitated the urinary excretion of systemic Al.56,59. Birchall and Chappell 1989 suggested that the geographical association between Al and AD might result from the effect of dissolved Silicon in promoting the formation of aluminosilicate species within the gastrointestinal tract, thereby reducing the absorption of Al from both water and other dietary sources.

References:

- Andrasi E., Pali N., Molnar Z., Kosel S. (2005). Brain aluminium, magnesium and phosphorus contents of control and Alzheimer-diseased patients. *J Alzheimer Dis*, **7**(4):273-84.
- Barel A., Calomme M., Timchenko A., De Paepe K., Demeester N., Rogiers V., Clarys P., Vanden Berghe D. (2005). Effect of oral intake of choline-stabilized orthosilicic acid on skin, nails and hair in women with photodamaged skin. *Arch Dermatol Res*; **297**(4):147–53.
- Bellia J.P., Birchall J.D., Roberts N.B. (1996). The role of silicic acid in the renal excretion of aluminum. *Ann Clin Lab Sci,* **26**:227–33.
- Berlyne G.M., Adler A.J., Ferran N., Bennett S., Holt J. (1986). Silicon metabolism I: some aspects of renal silicon handling in normal man. *Nephron*, **43**:5–9.
- Birchall J.D., Chappell J.S. (1989). Aluminum, water chemistry, and Alzheimer's disease. *Lancet*; **1**:953.
- Calomme M., Geusens P., Demeester N., Behets G.J., D'Haese P., Sindambiwe J.B., Van Hoof V., Vanden Berghe D. (2006). Partial prevention of long-term femoral bone loss in aged ovariectomized rats supplemented with choline-stabilized orthosilicic acid. *Calcif Tissue Int*; **78**(4):227–32.
- Carlisle E.M. (1972). Silicon: an essential element for the chick. *Science*; **178**:619.
- Carlisle E.M. (1982). Silicon overdose in man. *Nutrition Reviews*; **40**:208–9.
- Charnot Y., Pérès G. (1971). Modification de l'absorption et du métabolisme tissulaire du silicium en relation âvec l'age, le sehe et diverses glandes endocrines. *Lyon Medicine*; **13**:85.
- Dobbie J.W., Smith M.J.B. (1986). Urinary and serum silicon in normal and uraemic individuals. In: Evered, D.; O'Connor, M., editors. Silicon Biochemistry, Ciba Foundation Symposium 121. John Wiley and Sons Ltd.; Chichester: pp. 194-208.
- Exley C. (1998). Silicon in life: a bioinorganic solution to bioorganic essentiality. *Journal of Inorganic Biochemistry*; **69**:139–44.
- Gillette-Guyonnet S., Andrieu S., Nourhashemi F., De la Guéronnière V., Grandjean H., Vellas B. (2005). Cognitive impairment and composition of drinking water in women: findings of the EPIDOS Study. *Am J Clin Nutr* ; **81**(4):897-902.
- Iler R.K. (1979). Solubility, polymerisation, colloid and surface properties, and biochemistry. In : The chemistry of silica. John Wiley & Sons; New York.
- Jugdaohsingh R. (2007). Silicon and bone health. *J Nutr Health Aging*, **11**(2):99–110.
- Jugdaohsingh R., Anderson S.H., Tucker K.L., Elliott H., Kiel D.P., Thompson R.P.H. (2002). Dietary silicon intake and absorption. *American Journal of Clinical Nutrition*; **75**:887–93.
- Jugdaohsingh R., Tucker K.L., Qiao N., Cupples L.A., Kiel D.P., Powell J.J. (2004). Silicon intake is a major dietary determinant of bone mineral density in men and pre-menopausal women of the Framingham Offspring Cohort. *Journal Bone and Mineral Research*; **19**:297–307.

- Kawahara M. (2005). Effects of aluminum on the nervous system and its possible link with neurodegenerative diseases. *J Alzheimer's Dis*; **8**(2):171-81.
- Nielsen F.H. (1991). Nutritional requirements for boron, silicon, vanadium, nickel, and arsenic: current knowledge and speculation. *The FASEB Journal*; **5**:2661–67.
- Reffitt D.M., Jugdaohsingh R., Thompson R.P., Powell J.J. (1999). Silicic acid: its gastrointestinal uptake and urinary excretion in man and effects on aluminium excretion. *J Inorg Biochem*;**76**:141–47.
- Sjöberg S. (1996). Silica in aqueous environments. *Journal of Non-Crystalline Solids*; **196**:51–7.
- Smorgon C., Mari E., Atti A.R, Dalla Nora E., Zamboni P.F., Calzoni F., Passaro A., Fellin R. (2004). Trace elements and cognitive impairment: an elderly cohort study. *Arch Gerontol Geriatr Suppl*; **9**:393-402.
- Spector T.D., Calomme M.R., Anderson S., Swaminathan R., Jugdaohsingh R., Vanden-Berge D.A., Powell J.J. (2005). Effect of bone turnover and BMD of low dose oral silicon as an adjunct to calcium/vitamin D3 in a randomized placebo-controlled trial. *Journal of Bone Mineral Research*; **20**:S172.
- Walton J.R. (2006). Aluminium in hippocampal neurons from humans with Alzheimer's disease. *Neurotoxicology*; **27**(3):385-94.

Chapter 31
Silver (Ag)

Gea Oliveri Conti.

31.1 ENVIRONMENTAL EFFECT

Silver, atomic number 47, is a rare naturally occurring metal often found in association with other elements like Lead. Silver is a metal nearly white, lustrous, soft, very ductile, malleable and it is an excellent conductor of heat and electricity. It is not a chemically active metal but it can form silver salts. It has the highest electrical conductivity of all metals but it is very expensive.

The principal uses of silver, especially of silver nitrate, since the early days were the photography for its sensibility to light. The major applications of silver are photography, the electrical and electronic industries and for domestic uses as cutlery and jewellery (ATSDR, 1990).

Other applications are in dentistry and in high-capacity zinc long-life batteries.

Silver emissions in the biosphere are various and some of these are of anthropogenic in nature as during smelting operations, manufacture and disposal of certain photographic and electrical supplies and coal combustion. There are also natural sources of silver. The inputs of this metal affect atmosphere, water, and land. Silver is very stable in water and its level in soil is not usually high except in mineral-rich areas (ATSDR, 1990).Silver in rivers, lakes, and estuaries shows levels of about 0.01 µg/L in unpolluted areas and of 0.01–0.1 µg/L in urban and industrialized areas.

Silver concentrations reported prior to the implementation of specific ultra-clean metal sampling (coming out in 1980s) should be treated with extreme caution. It is important to note that environmental levels of silver have only a small portion of the total silver in each of the environmental compartments biologically available. Plants can absorb silver and measured levels come in the range 0.03-0.5 ppm.

However the ability to accumulate silver varies widely between species. Marine organisms are known for their bioconcentration capacity. Milligrams of silver per kilogram fresh weight organism divided by milligrams of silver per litre of medium give ratios of 210 in diatoms, 240 in brown algae, 330 in mussels, 2300 in scallops, and 18700 in oysters. These values represent uptake of bioavailable silver in animals in laboratory experiments. The bioconcentration however does not necessarily lead to adverse effects, the eucariotic bioconcentration however does not necessarily lead to adverse effects, in fact, at concentrations of silver normally assessed in the environment not are associated to the food-chain biomagnifications (Berthet et al., 1992). Increased biota silver concentrations occur, instead, in the vicinities of sewage outfalls, electroplating plants and mine waste sites.

Free silver ion is lethal to representative species of sensitive aquatic plants, invertebrates, and teleosts at nominal water concentrations of 1–5 µg/litre. Adverse effects occur on trout development at concentrations as low as 0.17 µg/litre and on phytoplankton species composition and succession at 0.3–0.6 µg/litre.

Silver accumulation by terrestrial plants from soils is very low. Silver is harmful to animal, including humans, at concentrations as low as 100 mg total silver/litre in drinking-water or 200 mg total silver/kg in diets (ATSDR, 1990). Sensitive laboratory mammals were adversely affected by total silver concentrations as low as 250 μg/litre in drinking-water with brain disease, by concentrations of 6 mg/kg in diet with high accumulations in kidneys and liver and decreased functionality, and finally concentrations of 13.9 mg/kg body weight were related to lethality.

31.2 EFFECT ON HUMAN HEALTH

Silver soluble salts, specially $AgNO_3$, are lethal in concentrations of up to 2g. Silver compounds can be slowly absorbed by body tissues with bluish or blackish skin pigmentation known as argyria (Wadhera and Fung, 2005). Silver may cause severe corneal injury and it may cause skin irritation. If the contact with skin is repeated and prolonged may cause allergic dermatitis. Exposure to high concentrations of vapours may cause dizziness, breathing difficulty, headaches or respiratory irritation. Extremely high concentrations may cause drowsiness, staggering, confusion, unconsciousness, coma or death (Drake and Hazelwood, 2005). Ingestion of silver may cause stomach discomfort, nausea, vomiting, diarrhoea, and narcosis. Aspiration of material into lungs or if vomiting occurs can cause chemical pneumonitis that can be fatal (Drake and Hazelwood, 2005). Health effect of silver on human body are listed in Table 31.1.
A reference dose (RfD) of 0.005 mg/kg/day for subchronic and chronic exposure was calculated from a lowest-observed-adverse-effect level (LOAEL) of 0.014 mg/kg/day for argyria observed in patients receiving i.v. injections of silver arsphenamine (Faust, 1992). Data are presently insufficient to derive a reference concentration (RfC) for silver (Faust, 1992).

Table 31.1 Silver effect in human body (ATSDR, 1990)
Kidney damage and reduction of nephron functionality
Eye damage
Lung damage
Liver damage
Brain and nervous damage
Cardiac abnormality

References:

- ATSDR (1990). Toxicological profile for silver. Atlanta, GA, US Department of Health and Human Services, Public Health Service, Agency for Toxic Substances and Disease Registry (TP-90-24).
- Berthet B., Amiard J., Amiard-Triquet C., Martoja M., Jeantet A. (1992). Bioaccumulation, toxicity and physico-chemical speciation of silver in bivalve molluscs: ecotoxicological and health consequences. *The Science of the Total Environment*, **125**:97–122.

- Drake P.L., Hazelwood K.J. (2005). Exposure-related health effects of silver and silver compounds: a review. *Ann. occup. Hyg*, **49**(7):575–85.
- Faust R.A. (1992). Toxicity summary of silver. Chemical Hazard Evaluation and Communication Group. Biomedical and Environmental Information Analysis Section, Health and Safety Research Division, Oak Ridge National Laboratory, Oak Ridge, Tennessee. 1-10.
- International Programme on Chemical Safety (IPCS). (2002). Concise International Chemical Assessment Document 44. Silver and silver compounds: environmental aspects. ISBN 92 4 153044 8 (NLM Classification: QV 297), ISSN 1020-6167. World Health Organization.
- Wadhera A., Fung M. (2005). Systemic argyria associated with ingestion of colloidal silver. *Dermatology Online Journal*, **11**(1):12.

Chapter 32
Sodium (Na)

Adriana Floridia, Sanjay Mishra.

32.1 ENVIRONMENTAL EFFECT

Sodium, atomic number 11, is an element in the periodic table that has the symbol Na (from Latin Natrium) and atomic number 11. It belongs to the group of alkali metals. It is a soft metal, waxy, silvery, reactive and very abundant in natural compounds. Sodium is the sixth most abundant element in the earth's crust, which contains 2.83% of sodium in all its forms. Sodium is, after chlorine, the second element most abundantly dissolved in sea water (1.05%). The most important salts of sodium present in nature are sodium chloride (rock salt), sodium carbonate (soda), sodium borate (borax), sodium nitrate and sodium sulfate (Lenntech, 2013). Sodium salts (e.g., sodium chloride) occurs widely in the environment, they are naturally present in drinking-water and in all food where may be also added during food processing. Food is the main source of daily exposure to sodium, primarily as sodium chloride (WHO, 2003). The sodium ion is ubiquitous in water. Most water supplies contain less than 20 mg of sodium per litre, but in some countries levels can exceed 250 mg/litre. Saline intrusion, mineral deposits, seawater spray, sewage effluents, and salt used in road deicing can all contribute significant quantities of sodium to water. In addition, water-treatment chemicals, such as sodium fluoride, sodium bicarbonate, and sodium hypochlorite, can together result in sodium levels as high as 30 mg/litre. Domestic water softeners can give levels of over 300 mg/l, but much lower ones are usually found (WHO, 2003).

32.2 EFFECT ON HUMAN HEALTH

Na^+ is the major ion in intracellular fluid and it is an essential ion for many body functions: maintenance of fluid balance, regulation of blood pressure and normal function of the nervous system (Meneton et al., 2005). The phisiological blood level of Na is 132 - 144 mEq/l (Farrell & Bower, 2003). Although it is generally agreed that sodium is essential to human life, there is no agreement on the minimum daily requirement. However, it has been estimated that a total daily intake of 120–400 mg will meet the daily needs of growing infants and young children, and 500 mg those of adults (WHO, 2003).

Hyponatremia has been referred to as "water intoxication," or low sodium level in the blood, this condition may result from excess water or fluid in the body (diluition effect) or by excess of water's losses without an adequate intake of sodium. Hyponatremia can be the result of chronic conditions such as kidney failure (excess fluid not be efficiently excreted) and congestive heart failure, in which excess fluid accumulates in the body.Hyponatremia furthermore sometimes can be associated with some diseases as a adrenal insufficiency, hypothyroidism, and cirrhosis of the liver (Meneton et al., 2005).

High levels of sodium chloride in the diet are linked to hypertension (Meneton et al., 2005). In general, Na salts are not acutely toxic because of the efficiency with which mature kidneys excrete sodium. However, acute effects and death have been reported following accidental overdoses of sodium chloride. Acute effects may include nausea, vomiting, convulsions, muscular twitching and rigidity, and cerebral and pulmonary edema (Farrel & Bower, 2003). Excessive salt intake seriously aggravates chronic congestive heart failure, and ill effects due to high levels of Na in drinking-water have been documented (Meneton et al., 2005; Schrier & Bansal, 2008). Although there is an association between hypertension and certain diseases, such as coronary heart disease, genetic differences in susceptibility, possibly protective minerals (potassium and calcium), and methodological weaknesses in experiments make it difficult to quantify the relationship, and sodium in drinking-water generally makes only a small contribution to total dietary sodium. No firm conclusions can therefore be drawn at present as to the importance of sodium in drinking-water and its possible association with disease (WHO, 2003).

References

- Farrell D.J., Bower L. (2003). Fatal water intoxication. *J Clin Pathol.* **56**(10):803–4.
- Meneton P., Jeunemaitre X., De Wardener HE., Macgregor G.A. (2005). Links Between Dietary Salt Intake, Renal Salt Handling, Blood Pressure, and Cardiovascular Diseases. *Physiol Rev*, **85**(2):679-715.
- Robertson J.S., Slattery J.A., Parker V. (1979). Water sodium, hypertension and mortality. *J Public Health*, **1**(4):295-300.
- Schrier R.W., Bansal S., (2008). Pulmonary Hypertension, Right Ventricular Failure, and Kidney: Different from Left Ventricular Failure? *CJASN*, **3**(5):1232-37.
- WHO. (World Health Organization) (2003). Sodium in Drinking-water. Background document for development of WHO. Guidelines for Drinking-water Quality. Geneva. WHO/SDE/WSH/03.04/15.
- http://www.lenntech.com/periodic/elements/ca.htm (accessed on 2 March 2012).

Chapter 33
Strontium (Sr)

Gea Oliveri Conti.

33.1 ENVIRONMENTAL EFFECT

Strontium (Sr), Atomic number 38, is a soft, silver-yellow, alkaline-earth metal. Strontium is a naturally occurring element found in rocks, soil, dust, coal, and oil. Strontium in the environment exists in four stable isotopes, ^{84}Sr (read as strontium eighty-four), ^{86}Sr, ^{87}Sr, ^{88}Sr.

Strontium can also exist as several radioactive isotopes; the most common is ^{90}Sr with a half life of 28 years (Hopkins et al., 1967), ^{90}Sr is formed in nuclear reactorsand during the explosion of nuclear weapons.

Strontium reacts vigorously with water and quickly tarnishes in air, so it must be stored out of contact with air and water. Due to its extreme reactivity to air, this element always naturally occurs combined with other elements and compounds. Strontium is used similarly to calcium and barium, but it is rarely employed for its higher cost. Principal uses of strontium compounds are in pyrotechnics, for the brilliant reds in fireworks and warning flares and in greases. Most strontium is used as the carbonate in special glass for television screens and visual display units. Water-soluble forms of strontium have the opportunity to pollute drinking water. Some strontium compounds dissolve in water and some strontium compounds in soil can dissolve in water and move deeper in the soil to underground water. Fortunately the concentrations of strontium in drinking water are usually quite low (ATSDR, 2004, State of Wisconsin, 2012).

Naturally occurring strontium is not radioactive and is either referred to as stable strontium or strontium.

Food and drinking water are the largest sources of exposure to strontium.

Most of the strontium in water is dissolved. Not much strontium ends up in drinking water. .Usually human activities (mainly by dumping waste directly in the water) cause the high strontium concentrations in water. High strontium concentrations can also be caused by settling of dust particles from air that have reacted with strontium particles from industrial processes (ATSDR, 2004, State of Wisconsin, 2012).

People can be exposed to strontium by air or dust, eating food, drinking water etc...

Strontium concentrations in the human body is related to strontium concentrations in food. Foodstuffs that contain significantly high concentrations of strontium are: grains, leafy vegetables and dairy products.

33.2 EFFECT ON HUMAN HEALTH

When strontium uptake is extremely high, it can cause disruption of bone development and dental changes; however, this effect can only occur when strontium uptake is in the range of thousands ppm.

Infants and young children who ingest too much strontium show a high probability to develope a condition called strontium rickets, that is a disease in which bones are thicker

and shorter than normal and may also be deformed (long bones in the legs, knee, etc...); in fact, calcium ions in the bone can be substituted by other bivalent metal ions, such as magnesium, strontium, barium, zinc, nickel, manganese, cobalt or lead (Aaseth et al., 2012).
These malformations can be very disabling. A turkish study on children who lived in a village with high levels of strontium in the soil reported signs of rickets including thinning of the cranium, delayed closure of the fontanelles, beadlike growths at the ends of the ribs, thickening of the wrists, and leg deformities.
Breastfeeding for 24 months was protective against these effects.
Strontium salts, and strontium ranelate inhibit bone resorption in rodents and monkeys and this effect is presumably due to direct effect on the osteoclasts. Likely, the balance between bone reabsorption and bone formation is affected only by strontium (Aaseth et al., 2012).
Strontium levels in food and drinking water are not high enough to be able to cause these effects (Xu et al., 1997).
Low levels of stable strontium has not been shown to affect adult health, but may cause disease in children. Children may be more susceptible than adults to the negative effects of radioactive strontium but a balanced diet with sufficient amounts of vitamin D, calcium and protein reduce the amount of strontium that is absorbed (Xu et al., 1997).
High levels of radioactive strontium can damage bone marrow and cause anemia and prevent the blood from clotting properly.
The only stable strontium compound that may cause cancer is strontium chromate, but this effect is due to the presence of chromium and not of strontium.
Radioactive strontium is much more harmful to health. When the uptake is very high, it may cause anaemia, and at extremely high concentrations it is even known to cause cancer by its mutagenic activity.
High levels of strontium can occur in water drawn from aquifers rich in strontium minerals. Radioactive strontium,does not occur in nature but is usually associated with nuclear power plant spills or nuclear weapons testing. Strontium-contaminated water has no taste or odor, the only way to assess if drinking water has elevated levels of strontium is to analyze by a certified testing laboratory.
EPA has set a Health Advisory Level of 4mg/l for stable strontium, indeed has set a limit of 8 picocuries/l for ^{90}Sr both in drinking water but these are not regulatory levels or legally enforceable standards (ATSDR, 2004; State of Wisconsin, 2011).

References:

- Aaseth J., Boivin G., Andersen O. (2012). Osteoporosis and trace elements – An overview. *Journal of Trace Elements in Medicine and Biology*, **26**:149–52.
- Agency for Toxic Substances and Disease Registry. ATSDR. (2004). Toxicological profile for Strontium. Atlanta, GA: U.S. Department of Health And Human Services, Public Health Service.
- Hopkins B.J., Casarett G.W., Tuttle L.W., Baxter R.C. (1967). Strontium-90 and intrauterine development in the rat. *J Embryol Exp Morphol.* **17**(3):583-91.
- State of Wisconsin. Strontium in Drinking Water. Department of Health Services, Division of Public Health with funds from the Agency for Toxic Substances and

- Disease Registry Public Health Service, U.S. Department of Health and Human Services. P-00292 (10/2011).
- Xu F., Zhang X., Liu J., Fan M. (1997). The effects of strontium in drinking water on growth and development of rat bone. *Wei Sheng Yan Jiu*, **26**(3):172-8.

Chapter 34
Thallium (Tl)

Michael R. Moore

34.1 ENVIRONMENTAL EFFECT

Thallium, atomic number 81, is a soft and pliable metal. It is colorless, odorless, and tasteless. Thallium has a similar ionic radii to potassium, which is one principle behind its toxicity. Thallium is a heavy metal that was serendipitously discovered by Sir William Crookes. The metal was isolated both by Crookes and Lamy in 1862. They had been expecting to isolate tellurium but instead found the new element thallium while trying to extract selenium from the by-products of sulfuric acid production. Crookes named the new element "thallium" from the Greek *thallos*, meaning "green shoot or twig" after the bright green spectral emission lines that identified the element.

In the past, thallium was used as a therapeutic agent to treat syphilis, gonorrhea, tuberculosis, and ringworm, and it was also used as a depilatory for excess hair. Thallium was also widely used as a rodenticide. Its use as a household rodenticide was banned in the United States in 1965 after multiple unintentional poisonings. Commercial use was banned a decade later. Thallium is used in the manufacture of electronic components, optical lenses, semiconductor materials, alloys, gamma radiation detection equipment, imitation jewelry, artist's paints, low temperature thermometers, and green fireworks. Thallium exposure may occur at smelters in the maintenance and cleaning of ducts and flues and through contamination of cocaine, heroin, and herbal products (ATSDR, 1992).

34.2 EFFECT ON HUMAN HEALTH

Thallium and its compounds are extremely toxic, and should be handled with great care. There are numerous recorded cases of fatal thallium poisoning. Contact with skin is dangerous, and adequate ventilation should be provided when melting this metal. Thallium(I) compounds have a high aqueous solubility and are readily absorbed through the skin. Exposure to them should not exceed 0.1 mg per m^2 of skin in an 8-hour time-weighted average (40-hour work week). Thallium is a suspected human carcinogen. Since thallium is a naturally-occurring element, it may be present in ambient waters in trace amounts. However, monitoring data indicate elevated thallium concentrations near industrial and commercial sources and hazardous waste sites. A survey of tap water from 3,834 homes in the United States detected thallium in 0.68% of samples at an average thallium concentration of 0.89 µg/L. Thallium was detected in 10% of urban stormwater runoff samples at concentrations ranging from 1 to 14 µg/L. Thallium has been measured in seawater at 0.01-14.00 µg/L. Water concentrations of thallium in rivers in the United States and Canada that receive mining operations effluents ranged from 0.7 to 88.3 µg/L (ATSDR, 1992).

For a long time thallium compounds were easily available as rat poison. This fact and that it is very water soluble and nearly tasteless led to frequent intoxications caused by accident or criminal intent (Emsley, 2006). Among the distinctive effects of thallium poisoning are loss of hair (which led to its initial use as a depilatory before its toxicity was properly appreciated) and damage to peripheral nerves (victims may experience a sensation of walking on hot coals), When present in aqueous solution as the univalent thallium(I) ion (Tl^{+}), it exhibits similarities with alkali metal cations, particularly potassium. It can thus enter the body via potassium uptake pathways. Other aspects of thallium's chemistry differ strongly from that of the alkali metals, such its high affinity for sulfur ligands. Thus this substitution disrupts many cellular processes

References:

- ATSDR, 1992. http://www.atsdr.cdc.gov/toxprofiles/tp.asp?id=309&tid=49
- Emsley J. (2006). "Thallium". The Elements of Murder: A History of Poison. *Oxford University Press*. pp. 326–327. ISBN 978-0-19-280600-0.

Chapter 35
Tin (Sn)

Cristina Mauceri, Ignatius Maduka.

35.1 ENVIRONMENTAL EFFECT

Tin, atomic number 50, in its most common form is a soft, silvery, white metal that is insoluble in water. Tin can combine with other chemicals to form various compounds. The most important inorganic compounds of tin are the oxides, chlorides, fluorides and halogenated sodium stannates and stannites. Inorganic tin compounds are found in small amounts in the Earth crust. Tin also can combine with carbon to form organotin compounds. Approximately 50% of the world production of tin is used for plating. Tin coatings are used for food containers and food-processing equipment. Tin is also used in alloys, such as solders, bronzes and pewters. Inorganic tin compounds are used as pigments in the ceramic and textile industry and for producing toothpaste, perfumes, soaps, food additives, and dyes.

There can be tin metal as well as inorganic and organic tin compounds in air, water, and soil near places where they are naturally present in rocks, mined, manufactured, or used. In general, organic tin compounds are from human-made sources and do not occur naturally in the environment. The time each tin compound stays in air, water, or soil differs from compound to compound (ATSDR). The background level of tin in air is about 0.01 μg/m^3, increasing to 0.3 μg/m^3 in urban areas and to 5 μg/m3 near industrial emissions. The concentration of tin in rivers, estuaries and oceans is generally less than 5 ng/l, but the use of organotin biocides can produce significantly higher concentrations. People can be exposed to tin by eating seafood from coastal waters or from contact with household products that contain organotin compounds, (polyurethane, plastic polymers, and silicon-coated baking parchment paper) but food, particularly canned food, represents the major route of human exposure to tin. Most natural foods contain tin in trace amounts, but concentrations are increased by the use of organotin pesticides and the storage of liquids in cans. In most unprocessed foods, tin levels are generally less than 1 μg/g. Higher concentrations are found in canned foods as a result of dissolution of tin coating or tin plate, the levels depending largely on the type and acidity of the food, the presence of oxidants, the duration and temperature of storage and the presence of air in the can headspace. Tin concentrations in foodstuffs frequently exceed 100 μg/g in unlacquered cans but are below 25 μg/g in lacquered cans. For the general population, drinking-water is not a significant source of tin (WHO, 2004).

35.2 EFFECT ON HUMAN HEALTH

Because inorganic tin compounds usually enter and leave your body rapidly after breathing or eating them, they do not usually cause harmful effects. However, some studies have shown that humans who swallowed large amounts of inorganic tin suffered stomachaches, anemia, and liver and kidney problems. Vomiting, diarrhea, fatigue and headache were often observed following the consumption of canned products (tin concentrations as low as

150 mg/kg in canned beverages and 250 mg/kg in other canned foods). Inhalation, oral, or dermal exposure to high amounts of certain organotin compounds for a short period of time can cause skin and eye irritation, respiratory irritation, gastrointestinal effects, and neurological problems in humans. Some neurological problems have persisted for years after the poisoning occurred. There are no data to indicate any adverse effects in humans associated with chronic exposure to tin (JECFA, 2000). JECFA concluded that there were insufficient data to establish an acute reference dose for inorganic tin but noted that concentrations of 150 mg/kg in canned beverages or 250 mg/kg in other canned foods may produce acute manifestations of gastric irritation in some individuals.

References:

- ATSDR. *Tin and tin compounds*. **1**:1-10.
- JECFA (Joint FAO/WHO Expert Committee on Food Additives). (2000) . Tin (addendum). *WHO Food Additives Series* No. 46.
- WHO. (2004). Inorganic Tin in Drinking-water: Background document for development of WHO Guidelines for Drinking-water Quality.; 1-5.

Chapter 36
Tungsten (W)

Adriana Floridia, Giovanni Arena.

36.1 ENVIRONMENTAL EFFECT

Tungsten (also known as Wolfram and represented by the letter W in the periodic table) atomic number 74, is a naturally occurring element that, in most environments, is solid. The average tungsten concentration in the Earth crust is about 0.006% (HSDB, 2009). In nature, it occurs in rocks and soil as mineral, but never as pure metal (EPA, 2012). Two kinds of tungsten-bearing mineral rocks, called wolframite and scheelite, are commercially mined. The mineral ore is processed to recover the tungsten and turn it into either chemical compounds or metal (ASTDR, 2005).

Based on its purity, the color of tungsten may range from white for the pure metal to steel-gray for the metal with impurities (EPA, 2012). Tungsten can be used as a pure metal or mixed with other metals to make alloys. Tungsten alloys tend to be strong and flexible, resist wear, and conduct electricity well.

Tungsten is found naturally in the Earth crust and is released into the atmosphere not only as a result of natural processes, such as the entrainment of dust particles and resuspension of the soil by the wind, but also through human activities. Tungsten and its alloys are used as light bulb filaments, as part of x-ray tubes where x-rays are formed, as catalyst to speed up chemical reactions, as component of steel in high-speed tools, in turbine blades, in phonographic needles, as welding electrodes, as gyroscope wheels, as counterbalance and fishing weights, in darts, and in golf club components. They can be used in bullets (as replacement for lead) and in arm or penetrators (as substitute for depleted uranium). Chemical compounds of tungsten are used for many purposes. Cemented tungsten carbide, a hard substance used to make grinding wheels and cutting or forming tools, is the most common tungsten compound. Other tungsten compounds are used in ceramic pigments, as fire retardant coatings for fabrics, and as fade-resistant dyes for fabrics (ASTDR, 2005).

Tungsten in water originates mainly from dissolution of tungsten from rocks and soil that water runs over and through. Only a very small fraction of tungsten in water originates from the settling of dust from air. Tungsten has not been detected in the vast majority of surface water and ground-waters. Some exceptions include areas near mines, natural deposits or industry where tungsten products of human-origin enter in waterways (ASTDR, 2005). Tungsten in drinking water has been detected, for example, in municipal water from Nevada (Seiler R.L. et al., 2005).

The general population may be exposed to low levels of tungsten by breathing air, drinking water, or eating food that contains tungsten. Exposure to tungsten above background levels may occur to the general population living near industries that process or use tungsten or its compounds, and to those living near hazardous sites that contain high concentrations of tungsten. The average ambient concentration of tungsten in air has been reported to be less than 10 ng/m^{3}of air (1 ng is 1 billionth of a gram). Cities have higher levels of tungsten in the air because tungsten is released from industry. Although very limited data are available, exposure to tungsten from air, drinking water, and food is expected to be insignificant. (ASTDR, 2005).

36.2 EFFECT ON HUMAN HEALTH

The toxicological profile of tungsten, including possible effects on living organisms and exposure pathways, remains rather sketchy, narrow and fragmentary. Regulation of tungsten, both in terms of environmental and occupational safety and health, is, at present, limited in comparison with other metals (Koutsospyros et al., 2006). EPA currently has no drinking water standard for tungsten (Art Fisher, EPA, 2012). Exposure limits have been established for soluble and insoluble tungsten compounds inworkplace atmospheric environments. Currently, the National Institute for Occupational Safety and Health (NIOSH) permissible exposure limits (PEL), expressed as time weighted average (TWA), stand at 1 and 5 mg/m^3 of air for soluble and insoluble compounds, respectively. (Koutsospyros et al., 2006, EPA, 2012).

Occupational exposure is considered the most common scenario for human exposure to tungsten and its compounds. Inhalation, ingestion, and dermal and eye contact are the possible exposure pathways. Occupational inhalation exposure to tungsten is known to affect the eyes, skin, respiratory system, and blood (EPA, 2012). Tungsten may cause irritation to eyes, skin and throat; diffuse interstitial pulmonary fibrosis; loss of appetite; nausea; cough; and changes in the blood (NIOSH, 2007).

References:

- Fisher A., Walker M., Powell P. Tungsten, Regional water quality program Southwest States and Pacific Islands.
- ATSDR. (2005). Toxicological Profile for Tungsten. Agency for Toxic Substances and Disease Registry.
- EPA. (2012). Technical Fact Sheet – Tungsten. U.S. Environmental Protection Agency.
- Hazardous Substances Data Bank (HSDB). (2009). Tungsten Compounds. http://t oxnet.nlm.nih.gov/cgibin/sis/htmlgen?HSDB (accessed on 12 September 2012).
- Koutsospyros A., Braida W., Christodoulatos C., Dermatas D., Strigul N. (2006). A review of tungsten:from environmental obscurity to scrutiny. *Journal of Hazardous Materials*, **136**:1-9.
- NIOSH. (2007). NIOSH Pocket Guide to Chemical Hazards: Tungsten. www.cdc.gov/niosh/docs/2005-149/pdfs/2005-149.pdf. (accessed on 12 September 2012).
- Seiler R.L., Stollenwerk K.G., Garbarino J.R. (2005). Factors controlling tungsten concentrations in ground water, Carson Desert, Nevada. *Applied Geochemistry*, **20**:423–41.

Chapter 37
Uranium (U) And Depleted Uranium (DU)

Gea Oliveri Conti.

37.1 ENVIRONMENTAL EFFECT

Uranium (U) is an unstable element that emits ionizing radiation. Natural uranium consists of a mixture of three radioactive isotopes that are identified by the mass uranium numbers ^{238}U (99.27% by mass), ^{235}U(0.72%) and ^{234}U(0.0054%).

Uranium is a naturally occurring, ubiquitous, heavy metal found in various chemical forms in all soils, rocks, seas and oceans. It is also present in drinking water and food.

On average, approximately 90 µg of uranium exist in the human body from normal intakes of water, food and air. It is also present in various organs of human body (see Figure 37.1) (WHO, 2003; WHO, 2005).

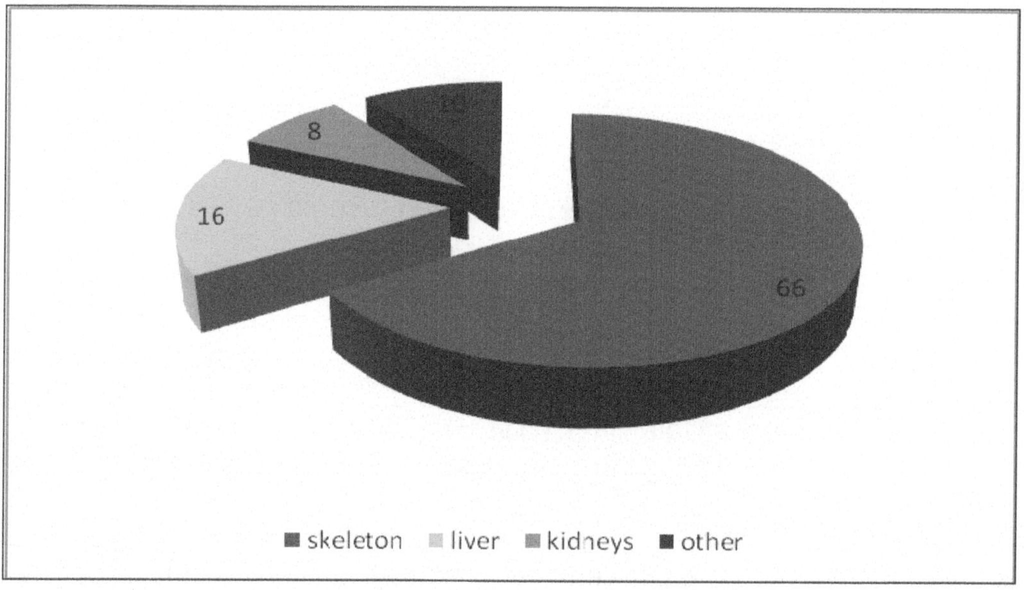

Fig. 37.1: Uranium content (%) in human body

Uranium emits radioactivity primarily in the form of alpha particles. This element exists in multiple forms, called "isotopes" that turn into other isotopes through a physical process known as "decay". Uranium is the "parent" substance or first element in the decay chain. U is a radioactive heavy metal that emits ionizing radiation of three types: alpha, beta and gamma (Tanaskovic et al., 2004; WHO, 2003) Radium and radon are, its "progeny" formed later in the decay chain. The unit of measurement for radioactivity is the becquerel (Bq), 1 Bq being the disintegration of one atom per second (UNEP, 2003).

Uranium (U) And Depleted Uranium (DU)

Uranium and radium are a metallic elements, while radon is a gas, all occur naturally in the environment (ARD-EHP-22, 2007). The radioactive half-life is the time it takes for a substance to lose (decay to) one-half of its radioactivity. The half–life of radon is only about four days whilst the half lives for uranium is approximately 4.5 billion years.

Uranium is used in the nuclear power industry, for military ammunition production, radiation shielding, and weighting balances for aircraft. Naturally occurring uranium is less radioactive than the uranium used in nuclear reactors, in which the ^{235}U isotope has been concentrated or "enriched" (ARD-EHP-22, 2007). Enriched natural uranium is used to make fuel for nuclear power plants; depleted uranium (DU) is the residue.from this process. DU is used as a counterbalance on helicopter rotors and airplane control surfaces, as a shield to protect against ionizing radiation, as a component of munitions to help them penetrate enemy armored vehicles, and as armor in some parts of military vehicles (ATSDR, 2011). Depleted uranium (DU) is a by-product from the process used to enrich natural uranium ore for use as fuel in nuclear reactors and nuclear weapons. It is distinguished from natural uranium by differing concentrations of some uranium isotopes. DU typically contains about 99.8% ^{238}U, 0.2% ^{235}U and 0.0006% ^{234}U by mass. For the same mass, DU has about 60% of the radioactivity of U.

We cannot perceive ionising radiation, so only analytical testing can determine if radionucleides are present in water (ARD-EHP-22, 2007; ATSDR, 1999). Uranium is almost as hard as steel and much denser than lead.

Only about 1-5 % of uranium is absorbed by ingestion. Food and drinking water are the primary sources of uranium intake for the general public. Vegetables such as potatoes, parsnips, turnips, and sweet potatoes contribute the highest amounts of uranium to the diet, the concentrations in these foods are directly related to the concentration of uranium in the soil where these foods are grown (ATSDR, 2011; WHO, 2005).

Because of its radioactivity, the amount of radioactive uranium decreases continuously but the half-life of .the isotope ^{238}U is 4.5 billion years. Therefore, the level of radioactivity does not change significantly.

A French study recently concluded that DU is fairly immobile in soil (UNEP, 2003). It was found that uranium had moved down the soil profile to a depth of 30 cm in approximately 30 years, and is held entirely within the 'A.1' fraction of the soil. The 'A' mineral horizon is defined as containing less than 17% organic carbon by mass that has formed at or near the soil surface in the zone of leaching or eluviations of organic materials in solution or suspension, or of maximum in situ accumulation of organic matter, or both. The thickness of the A.1 soil horizon depends on the local situation or geologic characteristics, but it does not usually exceed 30 cm (UNEP, 2003). Some experimental studies show that uranium moves 3000 times slower than the water that percolates through the soil.

About 10% of the uranium is able to move further down the soil profile and reach the underlying groundwater through the formation of U-humic acid colloids which aid in the transport of uranium. This could be the cause of somewhat elevated uranium levels, which reach 25 µg/L (ppb) in the groundwaters (UNEP, 2003).

The drinking water standard for tap waters set by the WHO is 2 µg/L, whereas the USEPA defined the standard of 30 µg/L, the Canadian Health has defined the standard of 10 µg/L (Weir, 2004) and the SSK in Germany has fixed it at 300 µg/L, though it is 2 µg/L in the territory of Hessen. EU drinking water directives give an indicative limit expressed as a dose of 0.1 mSv per year for radioactive substances in water.

The WHO guidelines for drinking-water quality, 2 µg/L of uranium, would apply to DU.

This corresponds to an average daily water intake with a uranium concentration of 100 µg/L (with 238U/234U in equilibrium).

Many bottled mineral waters have naturally high uranium contents of up to 100 μg/L. The guideline value of WHO for uranium in drinking-water is 0.03 mg/l (WHO, 2011).

37.2 EFFECT ON HUMAN HEALTH

Ionizing radiation can cause toxicity when the particles pass into or through the body at high speed. If a collision occurs with the molecules of living cells, they may be damaged.

Alpha radiation cannot pass through the dead outer layers of the skin.Uranium is thus a health risk only if taken into the body by ingestion or inhalation.

About 0.76–5% of uranium exposure occurs through the respiratory tract and 0.1–6% through the gastrointestinal tract. Water-soluble uranium compounds enter the bloodstream easily but a very small amount of uranium can be absorbed through the skin (ATSDR, 2011; WHO, 2005).

There are no known short-term effects from environmental exposure to uranium.

Natural uranium is weakly radioactive thus its more significant impact on human health is through toxicity like any other heavy metal, as it accumulates in the kidneys (WHO, 2005).

95% of uranium ingested or inhalated is not absorbed but it is excreted through faeces. Approximately 67% of the uranium absorbed into the blood will be excreted in the urine in 24 hours.

Between 0.2 and 2% of the uranium in food and water is absorbed by the gastrointestinal tract and soluble uranium compounds are more readily absorbed than insoluble ones.

Exposures to uranium are primarily associated with an increased risk of bone cancer, although cancer of other organs is also possible. There are no information on effects of uranium on human reproduction and development.

Animals exposed to uranium showed increased developmental effects but only at very high concentrations but it not likely to be found actually in drinking water (ARD-EHP-22, 2007; ATSDR,1999; ATSDR, 2011; WHO, 2005).

Concerns for possible illness to populations residing in conflict areas where depleted uranium munitions were used have raised many important environmental health questions (WHO, 2004; WHO, 2003). These populations can be exposed to depleted uranium in the same way as they are exposed to natural uranium, i.e. by inhalation, ingestion and dermal contact or injury by embedded fragments from munitions (WHO, 2004; WHO, 2003; Tanaskovic et al., 2004).

Inhalation is the most likely route of intake during or following the use of depleted uranium munitions in war or when depleted uranium is re-suspended in the atmosphere by wind or other forms of disturbance. Inhalation may also occur as a consequence of fire of tanks, of aircraft crash or of decontamination of vehicles from or near conflict areas (WHO, 2004; WHO, 2003).

Ingestion could occur in more sections of the population if their drinking water or food were contaminated with depleted uranium. Also, the ingestion of soil by children is considered another important pathway.

Kidneys and lungs are two important target organs of depleted uranium. Health consequences are determined by the physical and chemical nature of the depleted uranium, and by the level and duration of exposure.

Studies on workers exposed to uranium have reported some impairment of kidney function depending on the level of exposure.

However some studies showed that this illness may be only transient. Instead insoluble inhaled uranium particles as big as 1-10 μm retained in the lung may lead to irradiation damage of the lung and even lung cancer.

Direct contact with depleted uranium on the skin is unlikely to produce radiation-induced erythema (superficial inflammation of the skin) or other short term effects. Follow-up studies of veterans with embedded fragments in the tissue have shown detectable levels of depleted uranium in the urine, without important apparent health consequences.

The radiation dose to military personnel within an armoured vehicle is very unlikely to exceed the average annual external dose from natural background radiation from all sources (WHO, 2004; WHO, 2003).

The general public ingestion of soluble uranium compounds should not exceed the tolerable intake of 0.5 μg per kg of body weight per day. Inhalation of soluble or insoluble depleted uranium compounds by the public should not exceed 1 $\mu g/m^3$ in the respirable fraction.

Radiation dose limits are prescribed for exposures above natural background levels.For occupational exposure, the effective dose should not exceed 20 mSv per year averaged over five consecutive years, or an effective dose of 50 mSv in any single year. For exposure of the general public the effective dose should not exceed 1 mSv in a year; in special circumstances (WHO, 2005; WHO, 2004; WHO, 2003).

Gaps in knowledge exist and further research is recommended in war areas that would allow better health risk assessments to be made. In particular, studies are needed to clarify extent, reversibility and possible existence of thresholds for kidney damage in people exposed to depleted uranium (Occhigrossi & Nuccetelli, 2011; WHO, 2003).

References:

- ARD-EHP-22. (2007). Environmental Fact Sheet. Department of Environmental Services of New Hampshire . www. des.nh.gov. (accessed on 20 September 2012).
- ATSDR. (1999). Toxicological Profile for Uranium (Update). Agency for Toxic Substances and Disease Registry (ATSDR). Atlanta, GA. September, 1999.
- ATSDR. (2011). Public Health Statement. Uranium. Agency for Toxic Substances and Disease Registry (ATSDR). Atlanta, GA
- Occhigrossi A., Nuccetelli C. (2011). Monitoraggio biologico di militari e civili in zone contaminate da uranio impoverito: rassegna critica degli studi. ISSN: 1123-3117.
- United Nations Environment Programme. UNEP. (2003). Depleted Uranium in Bosnia and Herzegovina. Post-Conflict Environmental Assessment. Switzerland. ISBN 92-1-158619-4.
- Weir E. (2004). Uranium in drinking water, naturally. *JAMC*; **170**(6). DOI:10.1053/cmaj.1040214.
- WHO. World Health Organization. (2004). Depleted uranium: sources, exposure and health effects. http://www.who.int/ionizing_radiation/pub_meet/en/DU_Eng.pdf (accessed on 15 September 2012).

- WHO. World Health Organization (2003). Depleted uranium, Fact sheet N°257. http://www.who.int/mediacentre/factsheets/fs257/en/ (accessed on 15 September 2012).
- WHO. World Health Organization (2005). Uranium in Drinking-water. Background document for development of WHO Guidelines for Drinking-water Quality.
- WHO (World Health Organization). (2011). Guidelines for Drinking-water Quality. 4th ed. WHO, Geneva, Switzerland. ISBN 978 92 4 154815 1.
- Tanaskovic I., Pantelic G., Vuletic V., Javorina L., Eremic Savkovic M. (2004). Radioactivity of drinking waters from regions exposed to depleted uranium ammunition bombing. 11th International Congress of the International Radiation Protection Association (IRPA). Madrid, Spain, 23-28 May 2004. http://irpa11.irpa.net/pdfs/6c41.pdf (accessed on 20 September 2012).

Chapter 38
Vanadium (V)

Maria Fiore, Roberto Furnari.

38.1 ENVIRONMENTAL EFFECT

Vanadium, atomic number 23, is widely distributed in nature. It is abundant in rocks and soil (Sciacca et al., 1995). Evidence that environmental levels of vanadium are increasing has raised concern over the release of vanadium into the atmosphere from anthropogenic sources (Hope, 1994). Vanadium is a major trace metal in fossil fuels, and combustion of these materials provides a significant environmental source of this element. Vanadium in industrial particulate emissions was estimated to comprise about 53% of total atmosphere vanadium (Hope, 1994). Vanadium is poorly absorbed from a variety of foods but in sufficient quantities to be absorbed at detectable levels in many body tissues (French & Jones, 1993). The intake of vanadium via diet was reported to be 10 to 20 µg/d (Byrne & Kucera, 1991).

38.2 EFFECT ON HUMAN HEALTH

Vanadium is a trace element believed to be important for normal cell function and development. Various studies show that high vanadium exposure is contributing to many pathologies, see Table 38.1.

Table 38.1 Effect of Vanadium.

Bone effects	Increased bone growth in human and animals
Cardivascular effects	Hypo- and hypertension disease in human and animals, hypercholesterolemie in rat
Reproductive effects	Impaired sperm morphology and motility (rabbits and rats), reduced sperm number and decreased of fertility in mice and rat
Neurological effects	Damage by neurotoxicity of neuronal cells
Kidney effects	Damage of kidney in mice and rat
Immune effects	Reduction of immune response
Mutagenic effects	Increased of chromosomic aberrations
Weigh effects	Lowered BMI in woman, weight loss in rat
Developmental effects	Death of cardiomyocytes in rats fetuses, enhanced teratogenicity
Glucose balance	Decreased intake of water and generally of fluid, increased glucose tolerance

Numerous *in vitro* and *in vivo* studies have shown that vanadium has insulin-like effects in liver, skeletal muscle, and adipose tissue (Shechter & Shisheva, 1993; Verma et al., 1998; Badmaev et al., 1999; Thompson, 1999). It is present in all tissues, but its exact role in glucose homeostasis in man has yet to be established. Vanadium has the ability to imitate the action of insulin (Nechay, 1984; Nechay et al, 1986). Oral administration of inorganic vanadium salts has shown antidiabetic activity in vitro (Tolman et al., 1979), in vivo (Dubyak & Kleinzeller, 1980) and even in patients (Tamura et al., 1983).
One of the main mechanisms of vanadium action is through the generation of reactive oxygen species (ROS), by the activation of mitogen activated protein kinases (MAPKs) that will result in the phosphorylation of downstream enzymes such as ERK1/2, resulting in the activation or deactivation of diverse signaling pathways. Vanadium could also damage genetic material and induce hematotoxicity (Fortoul et al., 2008; Gonz´alez-Villalva et al., 2010), immune toxicity (Pi˜n´on-Zarate et al., 2007), hepatotoxicity (Avila-Costa & Fortoul, 2007), lung toxicity (Bonner et al., 2000; Knecht et al, 1985, Fortoul et al., 2007), neurotoxicity (Avila-Costa et al., 2004; Avila-Costa et al., 2005; Avila-Costa et al., 2006), and reproductive toxicity (Chandra et al., 2007, Arag´on et al.,2005; Fortoul et al., 2007). All these organs and systems are altered by vanadium inhalation. (Fortoul et al., 2007). DNA damage is induced by exogenous agents such as environmental pollutants (Taylor, 1994) or endogenous ones (free radicals) (Lindahl & Wood, 1999). Genetic damage mechanisms include oxidation of nitrogenous bases or sugar residues; cross-links or DNA strand breaks (Valko et al., 2006). Hematotoxicity include alterations in erythrocytes such as anemia, changes in hemoglobin and hematocrit concentrations (Zaporowska & Wasilewski, 1992), an increment in osmotic red blood cells fragility (Yang et al., 2003) modifications in white blood cells count (Zaporowska & Wasilewski, 1992), and an increase in platelets counts (thrombocytosis) and morphological changes as megaplatelets..
The Liver is also a target for inhaled vanadium. Increased lipid peroxidation is present during the exposure to this element demonstrating that the hepatocytes' cell membrane is altered by oxidative damage.
Vanadium exposure produces different effects on spleen such as splenomegaly, increase in megacaryocytes number and size, and the enlargement of the size of the white pulp.
In humans adverse effects of vanadium in the lung have been associated to exposure to suspended particles in polluted air. These studies report that inhalation exposure results in impaired lung function, influx of inflammatory cells in bronchiolar lavage, and fibrotic changes in the lungs (Bonner et al, 2000; Knecht et al, 1985).
The damage to the cytoskeleton of testicular cells produced by vanadium implies changes in testicular functions leading to infertility.
The central nervous system (CNS) exhibits a great vulnerability to oxidative stress damage, vanadium induced functional, cytological, and ultrastructural alterations in the olfactory bulb, motor cortex, hippocampus, striatum, substantia nigra, and ependyma (Avila-Costa & Fortoul, 2007).

References:

- Arag´on M.A., Ayala M.E., Fortoul T.I., Bizarro P., Altamirano-Lozano M. (2005). Vanadium induced ultrastructural changes and apoptosis in male germ cells. *Reproductive Toxicology*, **20**(1):127–34.

Vanadium (V)

- Avila-Costa M.R., Montiel Flores E.D., Col´ın-Barenque L., Ordoñez J.L., Gutiérrez A.L., Niño-Cabrera H.G., Mussali-Galante P., Fortoul T.I. (2004). Nigrostriatal
- modifications alter vanadium inhalation and inmunocytochemical and cytological approach. *Neurochemical Research*, **29**(7):1365–69.
- Avila-Costa M.R., Col´ın-Barenque L., Zepeda-Rodr´ıguez A., Antuna S.B., Saldivar O. L., Espejel-Maya G., Mussali-Galante P., del Carmen Avila-Casado M., Reyes-Olivera A., Anaya-Martinez V., Fortoul T.I. (2005). Ependymal epithelium disruption after vanadiumpentoxide minhalation: a mice experimental model. *Neuroscience Letters*, **381**(1-2):21–5.
- Avila-Costa M.R., Fortoul T.I., Ni˜no-Cabrera G., Colín-Barenque L., Bizarro-Nevares P., Gutiérrez-Valdez A.L., Ordóñez-Librado J.L., Rodríguez-Lara V., Mussali-Galante P., Díaz-Bech P., Anaya-Martínez V. (2006). Hippocampal cell alterations induced by the inhalation of vanadium pentoxide(V_2O_5) promote memory deterioration. *Neurotoxicology*, **27**:1007–12.
- Avila-Costa M.R. & Fortoul T.I. (2007). Vanadium and the liver, in Vanadium: Its Impact on Health, T. I. Fortoul and M. R. Avila-Costa, Eds., chapter 8, pp. 57–74, Nova Science, New York, NY, USA.
- Badmaev V., Prakash S., Majeed M. (1999). Vanadium: a review of its potential role in the fight against diabetes. *J Alternative Complementary Med.* **5**:273–91.
- Byrne A.R., Kucera J. (1991). New data on levels of vanadium in man and his diet. In Momcilovic B., ed. Trace elements in man and animals-7. Zagreb. Croatia, I99I :2S: 18-20.
- Bonner J.C., Rice A.B., Moomaw C.R., Morgan D.L. (2000). Airway fibrosis in rats induced by vanadium pentoxide. *American Journal of Physiology*, **278**(1):L209–16.
- Chandra A.K., Ghosh R., Chatterjee A., Sarkar M. (2007). Effects of vanadate on male rat reproductive tract histology, oxidative stress markers and androgenic enzyme activities. *Journal of Inorganic Biochemistry*, **101**(6):944–56.
- Dubyak G.R., Kleinzeller A. (1980). The insulin-mimetic effects of vanadate in isolated rat adipocytes. Dissociation from effects of vanadate a\a (Na'-K')ATPase inhibitor. *J Biol Chem*, **255**:5306-12.
- Fortoul T.I., Pi˜n´on-Z´arate G., D´ıaz-Bech M.E., González-Villalva A., Mussali-Galante P., Rodriguez-Lara V., Colin-Barenque L., Martinez-Pedraza M., Montaño L.F. (2008). Spleen and bone marrow megakaryocytes as targets for inhaled vanadium. *Histology and Histopathology*, **23**(11):1321–26.
- Fortoul T.I., Rodriguez-Lara V., Mussali-Galante P., Diaz- Bech P., Monta˜no L.F. (2007). Vanadium and respiratory system in Vanadium Its Impact on Health, Fortoul T.I. and Avila-Costa M.R., Eds., chapter 3, pp. 21–7, Nova Science, New York.
- Fortoul T.I., Bizarro-Nevares P., Acevedo-Nava S, Piñón-Zárate G., Rodríguez-Lara V., Colín-Barenque L., Mussali-Galante P., Avila-Casado Mdel C., Avila-Costa M.R., Saldivar-Osorio L. (2007). Ultrastructural findings in murine seminiferous tubules as a consequence of subchronic vanadium pentoxide inhalation. *Reproductive Toxicology*, **23**(4):588–92.
- French R.J., Jones P.J.H. (1993). Role of vanadium in nutrition: metabolism, essentiality and dietary considerations. *Life Sci*; 52.

- Gonz´alez-Villalva A.E., Falc´on-Rodr´ıguez C.I., Fortoul T.I. (2010). Signaling pathways involved in megakaryopoiesis. Gaceta M´edica de M´exico, **146**(2):136–43.
- Hope B.K. (1994). A global biogeochemical budget for vanadium. *Sci Total Environ*, **141**:1-10.
- Knecht E.A., Moorman W.J., Clark J.C., Lynch D.W., Lewis T.R. (1985). Pulmonary effects of acute vanadium pentoxide inhalation in monkeys. *American Review of Respiratory Disease*, **132**(6):1181–85.
- Yang X., Wang K., Lu J., Crans D.C. (2003). Membrane transport of vanadium compounds and the interaction with the erythrocyte membrane. *Coordination Chemistry Reviews*, **237**(1-2):103–11.
- Lindahl T., Wood R.D. (1999). Quality control by DNA repair. *Science*, **286**(5446):1897–1905.
- Mitchell E., Frisbie S., Sarkar B. (2011). Exposure to multiple metals from groundwater a global crisis: Geology, climate change, health effects, testing, and mitigation. *Metallomics*, **3**:874–908.
- Nechay BR. (1984). *Mechanisms of action of vanadium.* Annu Rev Pharmacol Toxicol.; **24**:501-24.
- Nechay B.R., Nanninga L.B., Nechay P.S.E. (1986). Vanadyl (IV) and vanadate (V) binding to selected endogenous phosphate. carboxyl, and amino ligands: calculations of cellular vanadium species distribution. *Arch Biochem Biophys*: **25**(1):128-38.
- Pi˜n´on-Zarate G., Rojas-Lemus M., Garcia-Zepeda E., Fortoul T.I. (2007). Metals and immune system. in Metals and Toxicological Implication in Health, Fortoul TI, Ed., chapter 5, pp. 43–64, Research Signpost.
- Sciacca S., De Naro Papa F., Catalano F., Mortellaro S., Bellassai F., Fiore M. (1995). Presenza e concentrazione di Vanadio. Un problema nuovo per le acque dell'Etna. Nota preliminare. *L'Igiene Moderna*, **103**:1-10.
- Shechter Y., Shisheva A. (1993). Vanadium salts and the future treatment of diabetes. *Endeavour.* **17**:27–31.
- Taylor J.S. (1994). Unravelling the molecular pathway from sunlight to skin cancer. *Accounts of Chemical Research*, **27**(3):76–82.
- Tamura S., Brown T.A., Dubler R.E., Larner J. (1983). Insulin-like effects of vanadate on adipocyte glycogen synthase and on phosphorylation of 95,000 dalton subunit of of insulin receptor. *Biochem Biophys Res Commun.* **I**(13):80-6.
- Thompson K. (1999). Vanadium and diabetes. *BioFactors.* **10**:43–51.
- Tolman E.L., Barris E., Bums M., Pansini A., Partridge R. (1979). Effects of vanadium on glucose metabolism in vitro. *Life Sci.*; **25**(1):159-64.
- Valko M., Rhodes C.J., Moncol J., Izakovic M., Mazur M. (2006). Free radicals, metals and antioxidants in oxidative stressinduced cancer. *Chemico-Biological Interactions*, **160**(1):1–40.
- Verma S., Cam M., McNeill J. (1998). Nutritional factors that can favorably influence the glucose/insulin system: vanadium. *J Am College Nutr.* **17**:11–8.
- Zaporowska H., Wasilewski W. (1992). Haematological effects of vanadiumon living organisms, *Comparative Biochemistry and Physiology C*, **102**(2):223–31.

Chapter 39
Zinc (Zn)

Maria Grazia D'Agati, Sanjay Mishra.

39.1 ENVIRONMENTAL EFFECT

Zinc, atomic number 30, is one of the most common elements, constituting 20 –200 ppm (by weight) of the Earth's crust. Zinc is found in the air, soil, and water and is present in all foods.

In its pure elemental (or metallic) form, zinc is a bluish-white, shiny metal. Powdered zinc is explosive and may burst into flames if stored in damp places. Metallic zinc is used in many ways in industry. A common use for zinc is to coat steel and iron as well as other metals to prevent rust and corrosion; this process is called galvanization. Metallic zinc is also mixed with other metals to form alloys such as brass and bronze. Metallic zinc is also used to make dry cell batteries. Zinc combines with other elements to form zinc compounds. Common zinc compounds found at hazardous waste sites include zinc chloride, zinc oxide, zinc sulfate, and zinc sulphide (ATSDR, 2005).

Zinc compounds are widely used in industry. Zinc sulfide and zinc oxide are used to make white paints, ceramics, and other products. Zinc oxide is also used in producing rubber. Zinc compounds, such as zinc acetate, zinc chloride, and zinc sulfate, are used in preserving wood and in manufacturing and dyeing fabrics. Zinc chloride is also the major ingredient in smoke from smoke bombs. Zinc compounds are used by the drug industry as ingredients in some common products, such as vitamin supplements, sun blocks, diaper rash ointments, deodorants, athlete's foot preparations, acne and poison ivy preparations, and antidandruff shampoos (ATSDR, 2005).

Zinc metal is not freely found in nature; rather it occurs in the +2 oxidation state primarily as various minerals such as sphalerite (zinc sulfide), smithsonite (zinc carbonate), and zincite (zinc oxide) (EPA, 2005).

Zinc enters air, water, and soil as a result of both natural processes and human activities. Most zinc enters the environment as the result of mining, purifying of zinc, lead, and cadmium ores, steel production, coal and wastes burning,. These activities can increase zinc levels in the atmosphere. Waste streams from zinc and other metal manufacturing and zinc chemical industries, domestic waste water, and run-off from soil containing zinc can discharge zinc into waterways. The level of zinc in soil increases mainly from disposal of zinc wastes from metal manufacturing industries and coal ash from electric utilities. Sludge and fertilizer also contribute to increased levels of zinc in the soil. In air, zinc is present mostly as fine dust particles. This dust eventually settles over land and water. Rain and snow aid in removing zinc from air (ATSDR, 2005).

In natural surface waters, the concentration of zinc is usually below 10 µg/litre, and in groundwaters, 10–40 µg/litre (Elinder et al, 1986). In tapwater, the zinc concentration can be much higher as a result of leaking from piping and fittings (Nriagu, 1980). The most corrosive waters are those with low pH, high carbon dioxide content, and low mineral salts content.

Exposure of the general population to zinc is primarily through ingestion. The average daily intake of zinc from food in humans is 5.2–16.2 mg zinc/day; assuming a 70-kg average body weight, this corresponds to 0.07–0.23 mg zinc/kg/day. Zinc is widespread in commonly consumed food, but tends to be higher in those of animal origin, particularly some sea foods. Meat products contain relatively high concentrations of zinc, whereas fruits and vegetables have relatively low concentrations. Other possible pathways for zinc exposure are water and air. Individuals involved in galvanizing, smelting, welding, or brass foundry operations are exposed to metallic zinc and zinc compounds (ATSDR, 2005).

39.2 EFFECT ON HUMAN HEALTH

Zinc is virtually nontoxic to living organisms. It is the only pre-, post-, and transitional element that is neither cytotoxic nor systematically toxic, nor is it carcinogenic, mutagenic, or teratogenic. Zinc is not stored in the body and excess intakes result in reduced absorption and increased excretion. (SCF, 2003). Acute toxicity is infrequent in humans, but arises from the ingestion of excessive amounts of zinc salts, either accidentally or deliberately as an emetic or dietary supplement. Vomiting usually occurs after the consumption of more than 500 mg of zinc sulphate (WHO, 2003). Ingestion or administration of even very large amounts of Zn and all its compounds does not have adverse long-term consequences (Stefanidou et al., 2006; Fosmire, 1990).

Zinc is a trace element essential for cell proliferation and differentiation. It is a structural constituent of many enzymes and proteins, including metabolic enzymes, transcription factors, and cellular signalling proteins (Beyersmann, 2002). Zinc is an important element in preventing free radical formation, in protecting biological structures from damage and in correcting the immune functions. Zinc deficiency increases the levels of lipid peroxidation in mitochondrial and microsomal membranes and the osmotic fragility of erythrocyte membranes. Zinc deficiency also produces impaired haemostasis due to defective platelet aggregation, a decrease in Tcell number and the response of T-lymphocytes to phytomitogens (Keen CL et al 1990; Tapiero H. et al 2003). Zinc deficiency produces growth retardation, anorexia, delayed sexual maturation, iron-deficiency anemia, and taste alterations (see Tab. 39.1) (Barceloux, 1999).

Genotoxicity studies conducted in a variety of test systems have failed to provide evidence for mutagenicity by zinc. However, there are indications of weak clastogenic effects following zinc exposure (ATSDR, 2005).

In suplementation of diets preparations intended to increase the zinc intake above that provided by the diet should not contain zinc levels that exceed dietary reference values, and should contain sufficient copper to ensure a ratio of zinc to copper of approximately 7, as is found in human milk because of the close relationship between these elements and the likelihood of Zn-induced Cu deficiency (WHO, 2001).

A number of epidemiological studies and reviews that involved or examined the efficacy of dietary zinc against prostate cancer have been reported (West et al., 1991; Platz et al, 2001). In contrast, Leitzmann et al. reported a significant increase in the risk of advanced stage prostate cancer by excessive zinc supplement (Leitzmann et al., 2003).

In conclusion, The Department of Health and Human Services and the International Agency for Research on Cancer (IARC) have not classified zinc for carcinogenicity.

Based on incomplete information from human and animal studies, the EPA has determined that zinc is not classifiable as carcinogenic to humans (ATSDR, 2005 (1)).

Zinc (Zn)

Table 39.1 Zinc effects on human body (WHO, 2003)	
Excess	**Deficiency**
Brain Lethargy Focal neuronal deficits	**Brain** Minor conductivity of nerve Psychiatric disorders Mental lethargy Disorder of sensorial organs
Respiratory apparatus Respiratory loss Metal fume fever	**Thymus** Atrophy of thymus
Gastrointestinal apparatus Vomiting Nausea Epigastric pain Diarrhoea	**Skin** Dermal and skin lesions Acrodermatidis
Prostate Increased risk of prostate cancer	**Reproductive system** Male and female infertility Retard of genital development
Systemic Symptoms Altered functionality of lymphocyte	**Systemic Symptoms** Increased risk of infection Immunologic disorders Growth retardation Increased lipoperoxidation Fragility of erythrocyte Anaemia

References:

- ATSDR Agency for Toxic Substances and Disease Registry (2005). Toxicological profile for zinc. U.S. Department Of Health And Human Services.
  ATSDR (1). Agency for Toxic Substances and Disease Registry (2005). Division of Toxicology ToxFAQsTM Zinc CAS # 7440-66-6.
- Barceloux D.G. (1999). *Zinc. Clin Toxicol*, **37**:279–92.
- Beyersmann D. (2002). Homeostasis and cellular functions of zinc. *Mat Wiss U Werkstofftech*, **33**:764–69.
- EPA U.S. Environmental Protection Agency (2005). Toxicological review of zinc and compounds. EPA/635/R-05/002.
- Elinder C.G. Zinc. In: Friberg L, Nordberg GF, Vouk VB, eds. (1986). Handbook on the toxicology of metals, 2nd ed. Amsterdam, Elsevier Science Publishers, 664-679.
- Fosmire G.J. (1990). Zinc toxicity. *Am. J. Clin. Nutr*, **51**:225-27.

- Keen C.L., Gershwin M.E. (1990). Zinc deficiency and immune function. *Annu Rev Nutr*, **10**:415–31.
- Leitzmann M.F., Stampfer M.J., Wu K., Colditz G.A., Willett W.C., Giovannucci E.L. (2003). Zinc supplement use and risk of prostate cancer. *J Natl Cancer Inst*, **95**:1004–7.
- Nriagu J.O., ed (1980). Zinc in the environment. Part I, Ecological cycling. New York, NY, John Wiley.
- Platz E.A., Helzlsouer K.J. (2001). Selenium, zinc, and prostate. *Epidemiol Rev*, **23**:93–101.
- SCF. (2003). Opinion of the Scientific Committee on Food on the tolerable upper intake level of zinc. European Commission.
- Stefanidou Æ.C., Maravelias Æ.A., Dona C., Spiliopoulou. (2006). Zinc: a multipurpose trace element. *Arch Toxicol*, **80**:1–9.
- Tapiero H., Tew K.D. (2003). Trace elements in human physiology and pathology: zinc and metallothioneins. *Biomed Pharmacother*, **57**:399–411.
- U.S. National Library of Medicine, Toxnet Database. Available online: http://toxnet.nlm.nih.gov (accessed January 21, 2011).
- West D.W., Slattery M.L., Robison L.M., French T.K., Mahoney A.W. (1991). Adult dietary intake and prostate cancer risk in Utah: a case-control study with special emphasis on aggressive tumors. *Cancer Causes Control*, **2**:85–94.
- WHO. World Health Organization (2003). Zinc in Drinking-water Background document for development of WHO Guidelines for Drinking-water Quality. Geneva.
- WHO 2001 Envirnmental Health Criteria 221

Chapter 40
Metals and Disinfection Treatment.

Margherita Ferrante, Maria Fiore.

A number of chemical methods, that utilize varioust metals, are used for water treatment at point-of-use or entry and for community water systems. These methods can be grouped into several main categories with respect to their purpose and the nature of the technology. The main categories to consider here are: (1) chemical pre-treatments by coagulation-flocculation, (2) hardness control and (3) chemical disinfection processes.

1) Chemical precipitation or coagulation and flocculation with various salts of aluminum (e.g., alum), iron, lime and other inorganic or organic chemicals are widely used processes to treat water for the removal of colloidal particles (turbidity) and microbes. Treatment of water by the addition of chemical coagulants and precipitants has been practiced since ancient times, even though the principles and physico-chemical mechanisms may not have been understood. Although alum and iron salts are the most widely used chemical coagulants for community drinking water treatment. Chemical coagulation-flocculation enhances the removal of colloidal particles by destabilizing them, chemically precipitating them and accumulating the precipitated material into larger "floc" particles that can be removed by gravity settling or filtering. Flocculation causes aggregation into even larger floc particles that enhances removal by gravity settling or filtration. Coagulation with aluminum or iron salts results in the formation of insoluble, positively charged aluminum or iron hydroxide (or polymeric aluminum- or iron-hydroxo complexes) that efficiently attracts negatively charged colloidal particles, including microbes. However, reductions in microbial contaminants as well as turbidity, and dissolved and colloidal organic matter are also achieved in this process. Despite the caveats and limitations, alum coagulation and precipitation to remove turbidity and other visible contaminants from water at the household level has been traditionally practiced for centuries in many parts of the world (Jahn & Dirar, 1979; Gupta & Chaudhuri, 1992; WHO–WSH, 2011). In some circumstances specific removal of anionic compounds like phosphate and arsenate can be achuieved with the use of Lanthanum- containing materials. (Hagheseresht et al., 2009).

2) Water hardness is the traditional measure of the capacity of water to react with soap. Hard water requires considerably more soap to produce a lather. Hard water often produces a noticeable deposit or precipitate (e.g. insoluble metals, soaps or salts) in containers. It is not caused by a single substance but by a variety of dissolved polyvalent metallic ions, predominantly calcium and magnesium cations, although other cations (e.g. aluminium, barium, iron, manganese, strontium and zinc) also contribute (WHO, 2011). Dissolved minerals contribute to the taste of drinking-water to varying degrees depending on the interaction of other factors, such as pH and alkalinity. Water with a hardness above approximately 200 mg/litre may cause scale deposition in the treatment works, distribution system and pipe work and tanks within buildings. It will also result in excessive soap consumption and subsequent “scum” formation. Soft water, with a hardness of less than 100 mg/litre, may, on the other hand, have a low buffering capacity and so be more

corrosive for water pipes. No health-based guideline value is proposed for hardness in drinking-water. Excessively hard water can also have corrosive tendencies. Naturally soft water can have aggressive properties towards the piping material through which it is distributed.Soft water that is not stabilized is corrosive and dissolves metal surfaces and pipes, resulting in the presence of certain heavy metals, such as cadmium, copper, lead and zinc, in drinking-water (National Research Council, 1977). To avoid the corrosion of piping materials, the water is normally conditioned or stabilized. Frequently, this involves increasing the alkalinity and/or adding corrosion-inhibiting substances (e.g. phosphates) in some form. Each divalent ion (**e**.g. Ca^{2+} or Mg^{2+}) in the water is replaced by two sodium ions. Softening also increases the sodium (and chloride) content of the drinking-water (WHO, 2011). Corrosion can be associated with health risks (from leachates such as lead, copper and other metals) and reduced lifespan of the distribution network and appliances (e.g. water heaters) **. C**entral water softening treatment usually involves lime (hydrated calcium oxide) or lime soda (lime plus sodium carbonate) and is commonly practised. These chemicals increase the precipitation of the calcium and magnesium carbonate, reducing the calcium hardness of the treated water. These waters are balanced to minimize post-precipitation of lime and should be stabilized as needed to control corrosivity. Water can also be softened by cation exchange, wherein the divalent cations (Ca^{2+}, Mg^{2+}, etc.) are replaced by sodium.

3) Chemical disinfection of drinking-water includes any chlorine- or iodine-based technologies, including chlorine dioxide, as well as bromine, ozone, other oxidants, strong acids and bases, ferrates and some antimicrobial metals (e.g. silver and copper). Disinfection with metals has been done with soluble, colloidal and larger solid (metallic) forms added to water (WHO, 2011). Silver has a long and intriguing history as an antibiotic primarily as a bacteriostatic agent in human health care. It has been developed for use in water purification, wound care, bone prostheses, reconstructive orthopaedic surgery, cardiac devices, catheters and surgical appliances. The bacteriostatic action of silver or silver compounds is proportional to the bioactive silver ion (Ag^{+}) released and its availability to interact with bacterial or fungal cell membranes. Silver metal and inorganic silver compounds ionize in the presence of water, body fluids or tissue exudates. The silver ion is biologically active and readily interacts with proteins, amino acid residues, free anions and receptors on mammalian and eukaryotic cell membranes (WHO-WSH, 2011).
Chronic ingestion or inhalation of silver preparations (especially colloidal silver) can lead to deposition of silver metal/silver sulphide particles in the skin (argyria), eye (argyrosis) and other organs. These are not life-threatening conditions but cosmetically undesirable. Silver is absorbed into the human body and enters the systemic circulation as a protein complex to be eliminated by the liver and kidneys. Silver metabolism is modulated by induction and binding to metallothioneins. This complex mitigates the cellular toxicity of silver and contributes to tissue repair. Silver allergy is a known contra-indication for using silver in medical devices or antibiotic textiles (Lansdown, 2006).
However, none of them are considered suitable for long-term use to disinfect drinking water for various important and valid reasons. Silver and copper are difficult to deliver to water and primarily only bacteriostatic. However, the extent to which silver alone inactivates microbes in water is limited, bacteria may develop silver resistance and many microbes, such as viruses, protozoan cysts and oocysts and bacterial spores, are not inactivated at silver concentrations employed for point-of-use drinking water treatment. Therefore, these agents are not recommended for routine disinfection of household water (WHO-WSH, 2011).

References:

- Gupta A., Chaudhuri M. (1992). Domestic water purification for developing countries. *Aqua* (Oxford). **45**(5):290-98.
- Hagheseresht et al., Wang, S.; Do D.D. (2009). A novel lanthanum-modified bentonite, Phoslock, for phosphate removal from wastewaters. *Applied Clay Science* **46**(4):369–375.
- Jahn S.A.A., Dirar H. (1979). Studies on Natural Water Coagulants in the Sudan with special Reference to Moringa oleifera seeds. *Water SA*, **5**(2):90-7.
- Lansdown A.B. (2006). Silver in health care: antimicrobial effects and safety in use. *Curr Probl Dermatol.*, **33**:17-34.
- National Research Council. (1977). Drinking water and health. Washington, DC, National Academy of Sciences.
- WHO, (2011). Hardness in Drinking-water Background document for development of WHO Guidelines for Drinking-water Quality. World Health Organization ,Genève.
- WHO-WSH. (2011). Water Sanitation and Health. Managing water in the home: accelerated health gains from improved water supply. http://www.who.int/water_sanitation_health/dwq/wsh0207/en/index6.html

Chapter 41
METALS REGULATIONS AND GUIDELINES OF SOME COUNTRY.

Margherita Ferrante, Gea Oliveri Conti.

The recent past has seen several initiatives to provide the same limits to the global control of the quality of drinking water but the goal is still very hard to reach.
USEPA, WHO, EC, Canadian and Australian guidelines, and today also Chinese guidelines, set standards for metal concentrations in drinking water but often limits for the same metal differ in these countries and some metals also not are fixed in many of them (eg. thallium, silver, vanadium, strontium, etc...) (see Table 41.1).
The International standard reference values for metals in drinking water are various and, though they are often based on WHO guidelines, they are extremely diversified in relationship to local issues and emerging problems which must be managed specifically (eg. Vanadium for volcanic countries, sodium for islands, arsenic for cultivated areas, etc...). This represent a very serious problem of public health for evaluating the global health effects of these substances on human health and because it represents an important bias in the global epidemiological evaluation.
Based on the data shown in Table 41.1 it is evident, in fact, that there is no uniformity in the standards limits for the tabulated countries.
Many metals are regulated on the basis of the WHO guidelines, however, some countries use the principle of precaution by adopting a lower limit (eg. Arsenic in the EU, Australia and New Zealand for Cadmium, Copper in Italy, Mercury in all country, etc...).
Some elements (Bismuth, Cobalt, Lithium, Potassium, Silicon, Tin, Tungsten) are not regulated by all guidelines because, WHO does not have sufficient data about their toxicity.
Calcium and potassium are not normally regulated because their presence in water is positive for the maintenance of health, however, EC guidelines fixed a limit of 100 mg/L for aesthetic purposes.
Sodium is regulated by WHO guideline (200 mg/L) and this value is uniformly applied to all countries evaluated with the exception of Australia and New Zealand (180mg/L), this limit has a dual purpose, namely the aesthetic control of the quality of drinking water and finally preventive measure for the maintenance of human health.
New compulsory standards were fixed for the Peoples Republic of China. On July 1, 2012, China's new national standards for drinking-water quality officially took effect. The new standards are much stricter than the ones that had been in use since 1985, China's number of regulated items exceeds those in developed countries and regions. All 106 items included were adopted from the WHO guidelines for drinking-water quality. These new standards were necessary because of the rapid development of Chinese industry and agriculture which has affected ground-water supplies that show increased pollution. This is a goal, because, this is the first time a developing country has implemented strict regulations on drinking-water quality and the same standards have been applied in rural and urban areas. However putting them into practice is still a open problem (Qu et al., 2012).

Table 41.1 Metals standards in Drinking Water limits in major country								
Metal (mg/L)	**USEPA**	**WHO**	**EC**	**Canada**	**Australia and New Zeland**	**Italy[+]**	**China**	**South Africa**
Aluminum	0.05-0.2	-	0.2	0.1 - 0.2	0.2	0.2	-	0.15
Antimony	0.006	0.020	0.005	0.006	0.003			
Arsenic	0.05	0.01	0.01	0.01	0.01	0.01	0.01	0.02
Barium	2	0.7	-	1	2	-	2	-
Beryllium	0.004	-	-	-	0.006	-	-	-
Bismuth	-	-	-	-	-	-	-	-
Boron	-	2.4	-	-	-	1	-	-
Calcium	-	75-200	100	-	-	-	-	-
Cadmium	0.005	0.003	0.005	0.005	0.002	0.005	0.005	0.005
Chromium	0.1	0.05	0.05	-	0.05	0.05	0.05	0.05
Cobalt	-	-	-	-	-	-	-	-
Copper	1	0.05	2	0.1	2	0.01	1	1
Iron	0.3	-	0.2	0.3	0.3	0.2	0.3	0.1
Lanthanum	-	-	-	-	0.002	-	-	-
Lead	0.015	0.01	0.01	0.01	0.01	0.01	0.05[§]	0.01
Lithium	-	-	-	-	-	-	-	-
Manganese	0.05	0.4	0.05	0.05	0.1	0.05	0.05	0.05
Mercury	0.002	0.006	0.001	0.001	0.001	0.001	0.002	0.001
Nickel	0.1	0.07	0.02	-	0.02	0.02	0.1	-
Potassium	-	-	-	-	-	-	-	-
Radon^	-	-	-	-	5	-	-	0.05
Selenium	0.05	0.01	0.01	0.01	0.01	0.01	0.01	0.05
Silicon	-	-	-	-	80[#]	-	-	-
Silver	0.1	-	-	-	0.1	-	0.05	-
Sodium	200°	200°	200°	200°	180°	200°	200°	200°
Strontium	-	-	-	-	7	-	-	-
Thallium	0.002	-	-	-	-	-	-	-
Tin	-	-	-	-	Unnecessary	-	-	-
Tungsten	-	-	-	-	-	-	-	-
Uranium	0.03	0.03	-	0.02	0.017	-	-	0.07
Vanadium	-	-	-	-	-	0.14[ç]	-	0.1
Zinc	5	-	-	5	3[#]	-	5	3

*: ^{226}Ra e ^{28}Ra in pCi/L;
^: mBq/L;
-: not regulated;
§: 0.01 from 25/12/2013;
°: aesthetic;
ç: Legislative Decree of 22 December 2011 as Directive 98/83/EC implementation
+: Legislative Decree n. 31 of 2 February 2001 – Italy
- Aesthetic

Metals regulations and guidelines of some country.

References:

- Australian Drinking Water Guidelines (2011). Australian Drinking Water Guidelines 6, 2011, Volume 1.
- Council Directive 98/83/EC of 3 November 1998 on the quality of water intended for human consumption. Directive 98/83/EC implementation on the quality of water intended for human consumption
- Decree of 22 December 2011 as Directive 98/83/EC implementation on the quality of water intended for human consumption. Change of parameter value for vanadium.
- Legislative Decree n. 31 of 2 February 2001 – Italy
- National Primary Drinking Water Regulations - EPA. http://water.epa.gov/drink/contaminants/index.cfm#List (accessed on 30 September 2012).
- National Secondary Drinking Water Regulations - EPA. http://water.epa.gov/drink/contaminants/index.cfm#List (accessed on 28 September 2012).
- Qu W., Zheng W., Wang S., Wang Y. (2012). China's new national standard for drinking water takes effect. **380**(3). www.thelancet.com (accessed on 28 September 2012).
- South African Water Quality Guidelines. Volume 1: Domestic Water Use Second Edition, 1996. http://www.dwaf.gov.za/IWQS/wq_guide/domestic.pdf. (accessed on 30 September 2012).
- http://eurex.europa.eu/LexUriServ/LexUriServ.do?uri=OJ:L:1998:330:0032:0054:IT:

IWA Publishing's authorised EU representative for General Product Safety Regulations is Diane D'Arras, 15 rue Duret, 75116 Paris, France, e-mail: safety@iwap.co.uk.

Printed and bound by CPI Group (UK) Ltd, Croydon, CR0 4YY
12/05/2026
02108868-0019